GENETIC ANALYSIS

GENETIC
ANALYSIS
Principles, Scope and Objectives

JOHN R. S. FINCHAM
PhD, ScD, FRS, FRSE
Division of Biological Sciences
University of Edinburgh

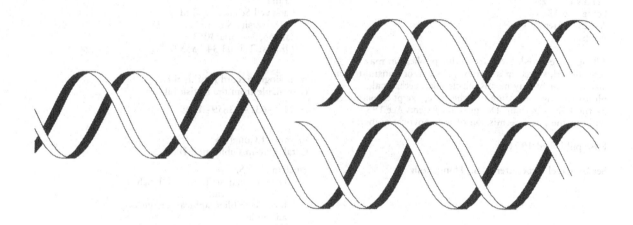

b

**Blackwell
Science**

© 1994 by
Blackwell Science Ltd
Editorial Offices:
Osney Mead, Oxford OX2 0EL
25 John Street, London WC1N 2BL
23 Ainslie Place, Edinburgh EH3 6AJ
238 Main Street, Cambridge
  Massachusetts 02142, USA
54 University Street, Carlton
  Victoria 3053, Australia

Other Editorial Offices:
Arnette Blackwell SA
1, rue de Lille
75007 Paris
France

Blackwell Wissenschafts-Verlag GmbH
Kurfürstendamm 57
10707 Berlin
Germany

Blackwell MZV
Feldgasse 13
A-1238 Wien
Austria

First published 1994

Set by Excel Typesetters Co., Hong Kong

DISTRIBUTORS

Marston Book Services Ltd
PO Box 87
Oxford OX2 0DT
(*Orders*: Tel: 0865 791155
        Fax: 0865 791927
        Telex: 837515)

USA
Blackwell Science, Inc.
238 Main Street
Cambridge, MA 02142
(*Orders*: Tel: 800 759-6102
        617 876-7000)

Canada
Oxford University Press
70 Wynford Drive
Don Mills
Ontario M3C 1J9
(*Orders*: Tel: 416 441-2941)

Australia
Blackwell Science Pty Ltd
54 University Street
Carlton, Victoria 3053
(*Orders*: Tel: 03 347-5552)

A catalogue record for this title
is available from the British Library

ISBN 0–632–03659–1

Library of Congress
Cataloging-in-Publication Data

Fincham, J.R.S.
    Genetic analysis/John R.S. Fincham.
        p.   cm.
    Includes bibliographical references
    and index.
    ISBN 0–632–03659–1
    1. Genetics—Technique.
    2. Nucleotide sequence.
    3. Genome mapping.
    I. Title.
    [DNLM: 1. Genetics. 2. Genome.
    3. Sequence Analysis, DNA.
    QH 430 F492ga 1994]
    QH441.F56 1993
    574.87′322—dc20

CONTENTS

4   The evolving concept of the gene, 95

5   Analysis of the whole genome, 129

Genetics increasingly dominates biological science. What began as a rather esoteric field, concerned with the mode of inheritance of minor variations and quirks, has become the main road to understanding how living organisms function. The great expansion of the scope of genetics is due to the reinforcement of classical genetic methodology with the newer molecular analysis.

The purpose of this book is to explain how modern genetics is actually done, to review what it tells us about the structure and functions of the genetic material, and to consider the extent to which all this new knowledge has solved, or looks like solving, the classical biological problems of variation and development. I have tried to show both the great power of genetic analysis and its present incompleteness – the enormous complexity of living systems will not yield to total analysis in the forseeable future.

In order to keep the book to a reasonable length it has been necessary to be very selective in the choice of material. I have thought it best to consider a limited number of examples in some detail so as to show the kinds of analysis that are possible, even at the expense of marginalizing other important topics. A good deal of the more detailed information is presented in the figures and their legends. The bibliography is mainly aimed at sup-

porting these specific examples, though I have also listed some more general sources of information.

This is not intended as a comprehensive genetics text-book. The emphasis is on 'mainline' eukaryotic organisms, with bacteria and viruses treated rather summarily, mainly in connection with their relevance to genetic manipulation. I have also concentrated on the problems that can be attacked through controlled investigations in laboratories, and have left aside questions of populational genetic change and evolution. These latter areas require different kinds of analysis and even different styles of thought.

J.R.S.F.
March 1994

Genetics can be said to have started with the rediscovery in 1900 of Mendel's rules, published in 1866, for the transmission of clearly defined distinguishing traits from one generation to the next. Originally formulated for peas, they were soon found to hold true for plants and animals generally. Clear-cut inherited differences could be attributed to hereditary units that soon came to be called genes. Initially, the genes were just symbols in a set of algebraic formulae, set up to describe the patterns of transmission of the differences. But, mainly as the result of the work of T.H. Morgan's school of fruit fly (*Drosophila melanogaster*) geneticists, the genes acquired a physical location. Each gene could be shown to reside at a specific position (locus) on one of the chromosomes of the cell nucleus.

This phase of genetic analysis is described in Chapter 1. It was based entirely on the natural breeding systems of the organisms concerned – not only the sexual cycle of plant and animals but also the very different modes of gene transfer found in bacteria. These natural systems have tended to be overshadowed in recent years by the 'genetic engineering' made possible by molecular technology, but they still provide the geneticist with an essential set of analytical tools. Moreover, they are what goes on all the time in the real world outside the laboratory, which in itself is a more than sufficient reason for knowing about them.

Until the advent of molecular biology, the genes remained intangible. The chromosome locus was merely the site of the determinant of a difference – usually a difference between the normal form of the organism and an aberrant variant, or *mutant*. For several decades after the formal establishment of the chromosome theory of Mendelian inheritance, there remained considerable scepticism about its general importance. The apparently trivial or freakish character of many of the inherited differences used in classical genetics encouraged some to argue that, even if genes existed, they were responsible only for superficial quirks superimposed on an essentially invariant species-specific substructure. This view became increasingly implausible as the number and range of known Mendelian variants was increased, especially through the use of radiation and other mutagenic treat-

INTRODUCTION: THE EXPANDING SCOPE OF GENETICS

ments. It became apparent that no feature of the organism was immune to the effects of mutation. All parts of the living system appeared to be dependent on the integrity of whatever it was that resided at the chromosome loci.

Nevertheless, it remained true for a long time that genes were detectable only in so far as they mutated. As late as the 1950s, at least one distinguished geneticist, Richard Goldschmidt, argued that the 'gene' was created by the mutation – that the 'mutant gene' was just a scar on the chromosome which, in its unscarred state, was an integrated whole, not divided into functionally distinct components.

What, more than anything else, gave solidity to the gene concept, was the detailed study, especially in *Drosophila*, of recurrent mutations at the same chromosome locus. The effects of such *allelic* mutations tended to be variations on a common theme. By the criterion of non-complementation (discussed in Chapter 2) they seemed to represent different degrees of defect in the same function of the organism. Whatever it was that resided at the chromosome locus, it clearly had a high degree of functional specificity.

As genetic mapping was pursued to a higher level of resolution, first in *Drosophila* and then in even more detail in micro-organisms, it was discovered that the functional units, or genes, were not indivisible, as had previously been assumed, but consisted of linear arrays of individually mutable sites. And as the effects of gene mutations were analysed to the biochemical level, it became clear in certain cases that mutations within a gene caused changes in the sequence of amino acids in a protein, and that the amino acid sequence of the protein was encoded in the linear structure of the gene.

This phase of genetic analysis, culminating in the concept of the gene as a repository of linearly encoded information for the synthesis of a specific macromolecule, will be described in Chapter 2. It proceeded against the background of the proof (by transformation experiments) of the genetic function of DNA and concurrently with the elucidation of the biochemical mechanisms whereby DNA is transcribed into RNA, which is then translated into protein structure.

Until the 1970s, however, it still remained true that genes were recognized only through the effects of their mutations. The gene as a molecular structure in itself remained elusive. The great change came with the development of new molecular techniques, first for detection (probing) of specific DNA sequences and then for amplifying and purifying genes or gene fragments as molecular clones. These revolutionary developments are described in Chapter 3. They gave physicochemical reality to the genes and showed them to be tracts of DNA. Genes at the DNA level conform to the expectations of classical genetics in that they reside within specific and relatively short chromosome segments but, as we shall see in Chapter 4, they turn out to display a complexity and variety of structure and function that could not previously have been imagined.

As a consequence of the molecular revolution, the agenda of genetic analysis has become radically changed. Formerly it started with genetic variation and attempted to define and map the genes responsible. The challenge was to account for inherited variation in terms of gene differences. Now genes of all kinds can be detected and mapped, whatever their functions and whether they mutate or not. Moreover, once genes have been isolated as DNA, alterations of any desired kind can be made in them to order outside the cell, and the altered genes reintroduced into the living system for observation of the consequences. Analysis of gene function is no longer necessarily dependent on random mutation.

With the powerful combination of classical and molecular methods of analysis, the scope and ambition of genetics is greatly increased. It becomes possible to pose questions about maximum objectives, and the extents to which they are likely to be realized. The final three chapters of this book address what may be considered as the three grand objectives of genetics.

The first, dealt with in Chapter 5, is the *genome project*, at present attracting a great deal of attention as applied to humans, but being pursued in several other species as well. The project is to make a complete molecular map and ultimately obtain complete DNA sequences of all the chromosomes, at least for one representative individual. For more complex organisms it is a formidable task, but feasible given sufficient resources. Once a complete DNA sequence has been obtained, it should be possible to recognize the genes, or potential genes, by computer-based scanning. Finding out what all the genes do will be much more difficult. It will be possible to deduce much, though not everything, about the structures of the proteins that they encode, but the only way of finding out what a given protein does for the organism is to see what difference it makes when it is lost or modified, either by random mutation or by DNA engineering. This brings us back to the analysis of genetic variants with which genetics started, but working from the gene to the effect rather than the other way round. This approach is feasible with yeast, far more difficult with the mouse, and possible with humans only at one remove – by using the mouse as a proxy or model.

A second grand objective of genetics, which is considered in Chapter 6, is to account for natural inherited variation in terms of defined gene differences at the molecular level. A comprehensive account is clearly out of the question for any species. It would involve the completion of a whole genome project for every individual. There is, however, some prospect of being able to define some of the gene differences that make relatively large contributions to populational variation, even if their individual effects are not completely clear-cut. Gene differences of small effect – and the effects grade all the way down to zero – are generally not worth pursuing. Nevertheless, if a particular variant form of a gene, whatever the magnitude of its effect, has already been defined in one individual, it can be relatively straightforward to screen for it in other individuals. It is all a question of knowing what to look for.

The final ambition of genetics, dealt with in the final chapter, is to account for the entire development of the organism in terms of the information encoded in the genome. The problem here lies in the sheer complexity of the operation. It would be quite unrealistic to think in terms of a number, however large, of parallel and separate connections between genes and traits of the developed organism. In reality, the system is a network, with a very large number of primary elements (the 10 000 or 100 000 genes and the proteins that they directly encode) forming innumerable cross-connections and loops. We cannot be sure that a complete description will ever be possible for any but the simplest organisms. But, as Chapter 7 attempts to show, we do begin to see some of the general principles of interaction that, with innumerable subtle variations, may account in principle for the development of living systems. The most fruitful approach is again through the study of heritable variation, but now tracing cause–effect relationships in both directions – from observable inherited differences back to the genes, and ('genetics in reverse') from defined changes in the genes out to the phenotype.

1

DISSECTING THE GENOME USING NATURAL GENETIC SYSTEMS

The eukaryotic system

Eukaryotes distinguished from prokaryotes

Gregor Mendel established his simple laws of heredity in peas, and T.H. Morgan and his colleagues used the fruit fly *Drosophila melanogaster* to show that Mendelian factors were located on physical structures, the chromosomes of the cell nucleus. The classical principles of genetics established by these pioneers were soon shown to apply to other higher plants and animals, to several kinds of fungi and algae, and to human beings. All of these very diverse organisms qualify as *eukaryotes*, a class defined by common features of cell organization which are quite different from those of the *prokaryotes* which, broadly speaking, are the bacteria and allied forms. Another characteristic of eukaryotes is their system of sexual

reproduction (see pp. 7–9) which is, as we shall see later in this chapter, very different from the modes of genetic mixing found in prokaryotes.

The formal rules governing hereditary transmission in eukaryotes were, as explained in this chapter, all worked out in the absence of any knowledge of deoxyribonucleic acid (DNA). But DNA is now known to be the material in which genetic information is encoded, and is central to all our thinking about genetic mechanisms. Although this chapter will not be so much concerned with molecules as later ones, it is well to put DNA in the centre of the stage from the start. Box 1.1 contains basic information about DNA structure and the way that specific sequences of DNA bases are faithfully transmitted through cycles of replication.

In eukaryotic, as opposed to prokaryotic cells, the bulk of the DNA is present in a cell *nucleus*, bounded by a nuclear membrane, often called the nuclear envelope. The nuclear DNA is divided between a number of microscopically visible bodies called chromosomes, literally 'coloured bodies' in reference to their staining by a range of microscopists' dyes. Another feature of eukaryotic cells is the presence of minor fractions of the DNA in internal membrane-bound organelles – namely *mitochondria*, the centres of energy-generating oxidative metabolism, and, in green plants, *chloroplasts*, the centres of photosynthesis. These organelles, both in the non-chromosomal organization of their DNA and in their sensitivity to antibiotics, seem to have bacterial affinities, and it has been conjectured that they may be the specialized and reduced descendants of prokaryotes that established symbiotic relationships with the

Box 1.1 DNA structure and replication

. . . .

DNA consists of chains of deoxyribonucleotide units, each containing one of four alternative bases — adenine or guanine (purines) or cytosine or thymine (pyrimidines).

Deoxyadenylic acid (deoxyadenosine 5'-monophosphate) as an example of deoxynucleotide structure:

The DNA chain has the 3'-hydroxyl of the deoxyribose of one unit condensed with the 5'-phosphate of the next:

Double-stranded DNA consists of two chains of opposite orientation (one 3'-to-5' and the other 5'-to-3') held together in a double-helix by specific hydrogen bonding between adenine–thymine and guanine–cytosine base-pairs:

----- Hydrogen bonds

continued

Box 1.1 *Continued*

Skeletal model of the double-helix (duplex):

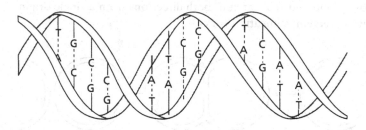

Note. The base sequence along one strand is the exact complement of, and therefore fixed by, the sequence along the other strand.

Replication of double-stranded DNA is catalysed by DNA polymerase and a number of accessory proteins. It proceeds by the progressive unwinding of the duplex at a replication fork and the synthesis of a new strand alongside each of the old ones. New synthesis is by successive addition of deoxynucleotide units to the 3′ ends of the growing chains. The deoxynucleotide to be added at each step is selected by the enzyme for correct base fit with the pre-existing strand, i.e., A opposite T, G opposite C. Because of the opposite 5′-3′ polarities of the two strands in the duplex, the new strands have to be synthesized in opposite directions. One can grow continuously with the moving replication fork and the other has to be synthesized 'backwards' in initially discontinuous fragments that are subsequently joined up.

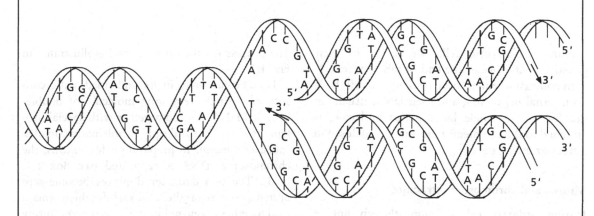

continued on p. 4

Box 1.1 *Continued*

The DNA of prokaryotes, plasmids and eukaryotic mitochondria and chloroplasts organelles is generally in the form of a
closed loop and replicated in both directions from a single origin,
with divergent replication forks.

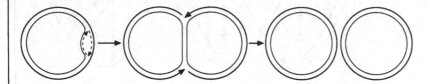

Chromosomal DNA replication proceeds from multiple origins,
of the order of 100–1000 per chromosome, with separate replication 'bubbles' eventually merging.

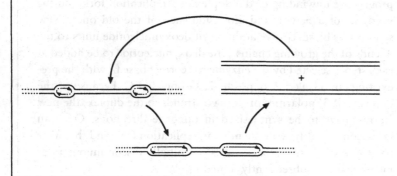

primitive eukaryotic cell. We return to their
genetic significance at the end of this chapter.

In contrast to eukaryotes, prokaryotic cells have
no internal organelles, and their DNA, usually in
the form of a single large folded molecule, is
attached to the cell membrane and not enclosed in
a nuclear envelope.

Mitosis and chromosome structure

During ordinary cell division (though not in
meiosis, see p. 9), each division of the cell is
preceded by the division of each of the chromosomes that constitute the cell nucleus. Each
daughter cell receives a nucleus with the same set
of chromosomes as was present in the mother cell.

The process is called mitosis and is illustrated in
Fig. 1.1.

The chromosome, in its undivided state, consists of a single very long molecule of double-
stranded DNA, packed together with special proteins to form a still deeply problematic complex
called *chromatin*. In preparation for mitosis, the
chromosomal DNA is replicated (see Box 1.1,
p. 3). The two daughter duplexes become separated into two parallel *chromatids* which remain
held together at one point, the *centromere*. During
the *prophase* stage the chromosomes become progressively condensed until, at the *metaphase* stage,
they are sufficiently compact to be visible as discrete, countable bodies under the microscope.

Metaphase is characterized by the appearance

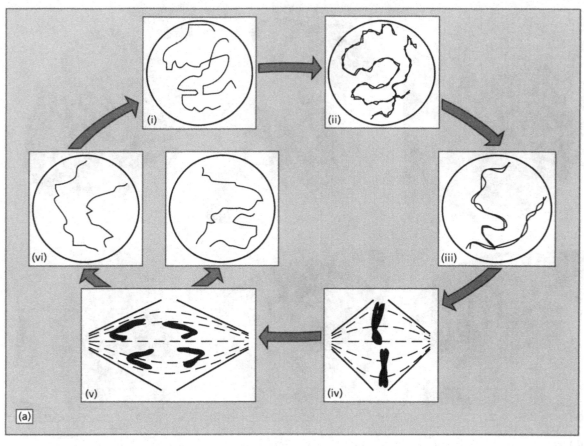

Fig. 1.1 (a) The mitotic cycle in a typical eukaryote. (i) Interphase before replication (G1); (ii) interphase after replication (G2); (iii) prophase, with chromosomes becoming contracted; (iv) metaphase, with chromosomes maximally contracted and aligned on the equator of the division spindle, the nuclear envelope having dissolved; (v) anaphase, with daughter chromosomes pulled apart by their centromeres to the spindle poles; (vi) telophase – nuclear envelopes form round daughter chromosome groups and chromosomes lose their compaction. For simplicity only two chromosomes are shown. The centrosomes (spindle pole bodies), which are seen in animals and fungi but not in flowering plants, are not shown here.

(*Continued overleaf.*)

of the *division spindle*, an array of parallel protein fibres with contractile properties, and the disappearance (except in fungi) of the nuclear envelope. The centromeres of the metaphase chromosomes become attached to spindle fibres and come to lie in a plane, the spindle equator, midway between the spindle poles. Metaphase is succeeded by *anaphase*, when each centromere splits into two daughter centromeres and the chromatids become separate daughter chromosomes. The daughter centromeres are pulled apart by their attached spindle fibres towards the spindle poles, with the flanking chromosome arms trailing

behind them. At *telophase*, the chromosomes lose their compact structure, and the two groups of daughter chromosomes become enclosed in nuclear envelopes to provide working nuclei for the two products of cell division.

It is at first sight difficult to believe that the relatively bulky metaphase chromatid could represent a single double-stranded DNA molecule. The quantity of DNA in the largest human chromosome, to take one example, would be about 8 cm in length if free of protein and fully extended. The chromosome at metaphase is of the order of 10 000 times shorter than that. Current

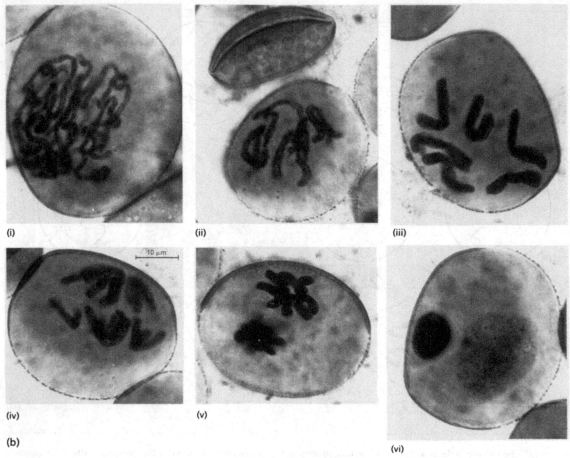

(i)   (ii)   (iii)

(iv)   (v)

(b)

(vi)

Fig. 1.1 (*continued*) (b) Stages of mitosis in pollen grains of *Tradescantia paludosa*, a plant species with very large chromosomes: (i) prophase; (ii) pro-metaphase; (iii) metaphase; (iv) anaphase; (v) early telophase; (vi) post-mitosis with two nuclei in the pollen grain. The more diffuse nucleus does not divide again, and the more condensed one divides in the pollen tube to form the two gamete nuclei (cf. Fig. 1.4).

understanding of how the necessary contraction and thickening is achieved is summarized in Fig. 1.2. The best understood part of chromatin structure is the first level of contraction – the formation of *nucleosomes* by the wrapping of the DNA around bead-like complexes of proteins of the *histone* class, and the helical packing of the nucleosome string to form a chromatin fibre. Further contraction is achieved by the attachment of the fibre in loops to a chromosome matrix, which itself may be helically folded as a final stage of compaction.

Metaphase is the stage at which chromosomes are most easily characterized and counted. Every

species has its own characteristic chromosome set, constant in number and form. The most important criteria used for distinguishing between chromosomes microscopically are: (i) length; (ii) position of the centromere, which may be central, off-centre or terminal; and (iii) the pattern of dark and light regions seen after certain special staining procedures (banding techniques, see p. 140).

Depending on the stage of the sexual life cycle, the cell nucleus usually contains either a single (*haploid*, n) set of chromosomes, with every one different, or a double (*diploid*, $2n$) set, with two chromosomes of each kind. Polyploid species, with

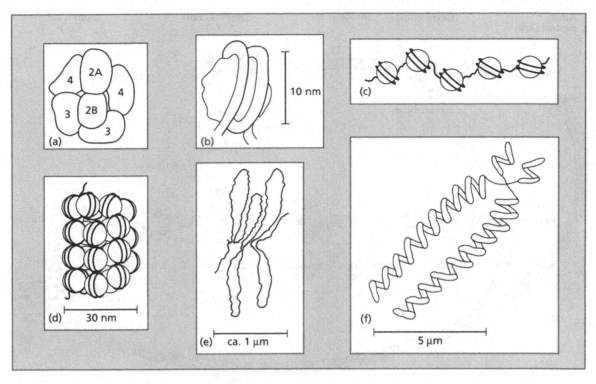

Fig. 1.2 Chromosome structure. (a) The structure of the protein core of the nucleosome, with two molecules each of histones 2A, 2B, 3 and 4 packed together to form a spherical protein core (one 2A and one 2B are out of sight behind the particle). (b) The winding of the DNA duplex around each nucleosome core. (c) The extended nucleosome string. (d) The packing of the nucleosome string to form the approximately 0.03 μm (30 nm, 300 Å) chromatin fibre. The compacted fibre is thought to be stabilized by the binding of a histone H1 molecule between each nucleosome and the next. After Widom & Klug (1985). (e) Looping of the chromatin fibre by its attachment at intervals to the nuclear matrix. At metaphase, shown here, the elements of the matrix coalesce to form a continuous linear chromosome scaffold. Drawn from the photograph of Marsden & Laemmli (1979). (f) Helical folding of the chromosome scaffold at metaphase (human chromosomes). Drawn from photographs of Boy de la Tour & Laemmli (1988).

$3n$, $4n$, $6n$, etc., chromosomes, are widespread and particularly important in plants, but are neglected in this book.

Alternation of haploid and diploid phases in the sexual cycle

The essence of sexual reproduction does not lie in the distinction between male and female – that is a common but not universal feature of sexually reproducing organisms – but rather in the alternation of haploid and diploid phases in the life cycle. The germ cells (*gametes*) – eggs and sperm in animals, ova and pollen tubes in flowering plants, morphologically undifferentiated cells in budding yeast – are haploid. The union of the germ cells leads to fusion of their nuclei (*karyogamy*) and initiates the diploid phase in which the parental haploid chromosome sets are present together. The transition back to haploid, which is essential for the next turn of the sexual cycle, occurs through the process of *meiosis*, described in the next section. Karyogamy and meiosis are the two key events that punctuate the sexual cycle.

Different groups of eukaryotes differ greatly in the relative duration and prominence of their haploid and diploid phases. In animals the only haploid cells are the gametes. Diploidy is maintained through many rounds of mitosis during the entire growth and development of the animal except in female oocytes and male spermatocytes, which undergo meiosis to form eggs and sperm,

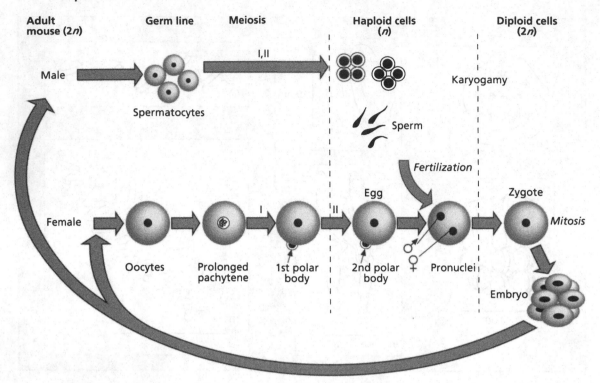

Fig. 1.3 The sexual life cycle of a mammal such as the mouse. The only haploid (*n*) cells are the immediate products of meiosis. Meiosis in the spermatocyte results in tetrads of spermatids that can all develop into sperm. Meiosis in the oocyte starts soon after the establishment of the female germ line, but is not completed until after fertilization by a sperm cell, which provides the haploid male pronucleus. The haploid female pronucleus is the only surviving product of female meiosis; one product of the first division and one product of the second division are discarded in the first and second polar bodies. Male and female pronuclei fuse to form the diploid (2*n*) zygote, which then undergoes cleavage to initiate embryonic development. In mammals the oocytes are stored with meiosis arrested at pachytene. *Drosophila* differs only in that female meiosis is not initiated at all until fertilization of the oocyte.

respectively. Figure 1.3 summarizes the sexual cycle in the mouse.

Flowering plants come close to the animal pattern in that (leaving aside polyploid species) their vegetative structures are entirely diploid. They differ from animals in a brief propagation of the haploid phase, following meiosis but preceding karyogamy, within the pollen tube and the embryo sac of the ovule (Fig. 1.4). Readers with a broader interest in plants may note that ferns, mosses and liverworts have more extensive haploid phases, amounting in ferns to the small and short-lived but free-living green prothallus and in mosses and liverworts to the whole green plant with the exception of the spore-bearing capsule, which is diploid.

Most fungi, including the experimentally im-

portant filamentous Ascomycetes, are haploid, with meiosis following immediately after karyogamy. But budding yeast, *Saccharomyces cerevisiae*, the fungus most prominently featured in this book, is unusual in being able to propagate itself vegetatively in either the haploid or the diploid phase (Fig. 1.5). Haploid cells are not distinguishably male and female, but exist in two self-incompatible but mutually compatible mating types. So long as the mating types are kept separate the haploid condition can be maintained, with the reservation that some strains, including most wild strains, are able to switch mating type. Diploid cultures are stable so long as they are well nourished, but are induced to undergo meiosis by starvation.

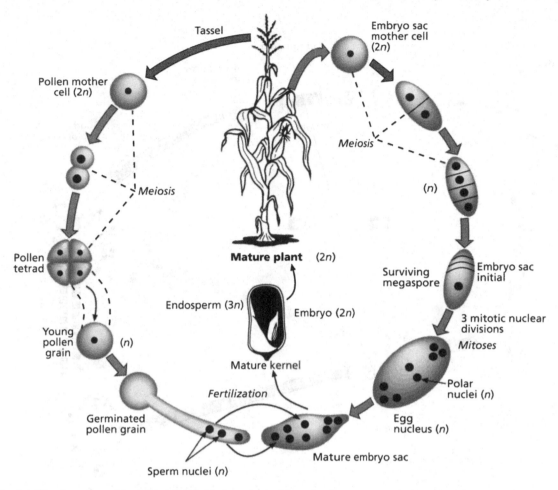

Fig. 1.4 The alternation of haploid and diploid phases in the life cycle of maize (*Zea mays*). Following meiosis, there are three rounds of haploid mitosis in the embryo sac before egg formation and, in the pollen grain and pollen tube, two haploid mitotic divisions. The endosperm, which provides a food store in the seed, is often triploid, the product of fusion of two embryo sac nuclei and one pollen tube nucleus. Based on Srb *et al.* (1965).

Meiosis and the rules of classical genetics

Meiosis

The process of meiosis, with some local variations some of which we shall deal with later, conforms to a standard pattern throughout the eukaryotic world. Its principal stages are illustrated in Fig. 1.6, and may be summarized as follows.

1 DNA replication occurs before the start of meiosis, and there is no further DNA replication (apart from some basic repair–replication associated with crossing over; see below) until after it is finished. The 2*n* chromosomes, notwithstanding the fact that initially they do not look double, start meiosis already replicated. They actually comprise four copies of the genome, which, through the two divisions of meiosis, are segregated from one another to provide one copy for each of the four meiotic products.

2 Pairs of matching (*homologous*) chromosomes become closely aligned point-for-point along their entire lengths. At the *pachytene* stage, when pairing (*synapsis*) is complete, alignment is so close that, even with the best resolution of the light microscope, there may be difficulty in seeing any space between the homologues. Electron micro-

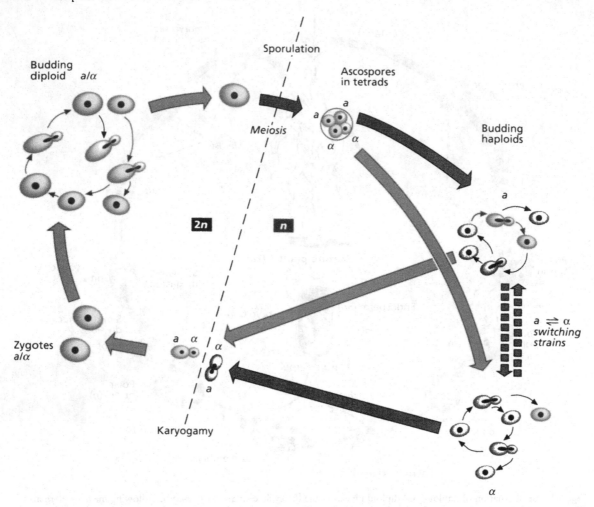

Fig. 1.5 The alternation of haploid and diploid phases in budding yeast, *Saccharomyces cerevisiae*. Haploid cells are of two alternative mating types *a* and α; all diploids are of genotype *a*/α and meiotic tetrads each contain 2 *a* and 2 α ascospores. In the presence of a switching gene *HO*, which is prevalent in the wild yeast, *a* and α mating types interconvert frequently by controlled DNA rearrangement, so any haploid quickly reverts to the diploid condition. The standard laboratory strains are homozygous for the non-switching allele *ho*, and so *a* and α strains can be propagated stably as haploids until brought together.

scopy reveals a protein ribbon, called the *synaptonemal complex*, running between each pair of synapsed chromosomes and consisting of a central element and two lateral elements, the latter apparently in direct contact with the chromosomes. The lateral elements bind to silver, permitting brilliantly clear electron microscopic visualization (Fig. 1.7).

3 The transition from pachytene to diplotene is marked by two striking visible changes. Firstly, the chromosomes can now be seen to be divided lengthwise into chromatids except at their centromeres, where the chromatids are still attached together. Secondly, the formerly synapsed chromosomes have largely separated, and are prevented from falling apart altogether by the presence between each pair of one or more cross-connections (*chiasmata* – chiasma in the singular). Each chiasma involves just one chromatid from each chromosome; the choice of chromatids for participation in any one chiasma appears to be a matter of chance. There is convincing micro-

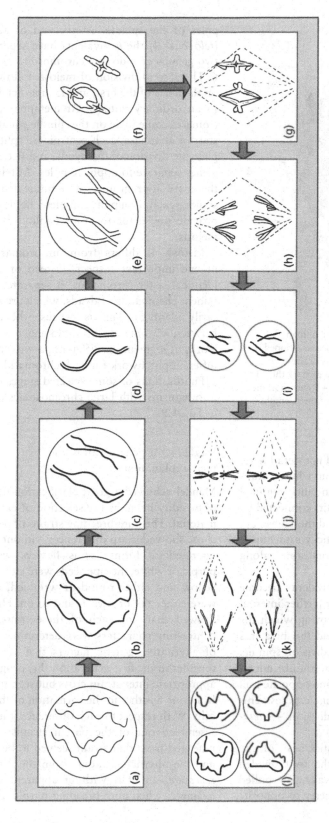

Fig. 1.6 Schematic representation of the key steps in typical meiosis. For simplicity, only two chromosome pairs are shown. (a and b) Premeiotic nucleus before and after DNA replication; $2n$ undivided chromosomes become $2n$ divided chromosomes (their internal division not yet visible) in process of pairing (synapsis). (c) Zygotene stage of prophase of the first meiotic division: homologous chromosomes somewhat contracted and more clearly visible. (d) Pachytene: synapsis complete – chromosomes somewhat contracted and more clearly visible. (e) Diplotene: chromosomes now visibly split into chromatids except at the centromeres; pairing lapses except at the chiasmata, one or more of which join each chromosome pair; at each chiasma two chromatids appear to cross between chromosomes. (f) Diakinesis: n pairs of joined chromosomes (bivalents) have become much more contracted. (g) Metaphase I: bivalents aligned on the spindle equator; each bivalent appears as if pulled towards opposite poles of the spindle by its centromeres, but is held together by its chiasma(ta). (h) Anaphase I: chiasmata are pulled apart; n dyads, each with an unsplit centromere, pass to each pole. (i) Telophase I (often very abbreviated): dyads become more diffuse, but contract again at prophase II (not shown). (j) Metaphase II: n dyads aligned on each of the two second division spindles. (k) Anaphase II: the centromeres split and the two chromosomes from each dyad disjoin to opposite poles. (l) Telophase II: four meiotic products each with n chromosomes.

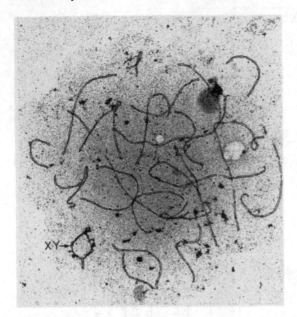

Fig. 1.7 Electron micrograph of a spread human spermatocyte at the pachytene stage of meiosis, stained with silver to show the paired lateral elements of the synaptonemal complex running between each of the 23 pairs of synapsed chromosomes. The X–Y chromosomes (arrowed) are paired along only a part of their lengths. Courtesy of Dr N.R. Davidson.

poles of the spindle. At the end of anaphase I (*telophase I*) the *n* bivalents have separated into two groups of *n* double chromosomes or *dyads*.

6 The second division of meiosis follows quickly on the heels of the first, and is superficially similar to an ordinary mitosis. At metaphase, the centromeres come to lie on the spindle equator; at the start of anaphase, each centromere splits to complete the division of the dyads, and sister chromosomes separate to opposite poles. At telophase II, there are four groups of *n* undivided chromosomes, each group a complete haploid set, to provide nuclei for the four haploid products of meiosis.

Meiosis II differs from an ordinary mitosis in one important respect. Instead of separating identical sister chromatids, it separates the component chromatids of dyads, which are not necessarily identical. This is because the crossovers between chromatids at diplotene create patchworks of segments of different parental origins – a different patchwork for each chromatid (Fig. 1.8).

Photographs of some selected stages of meiosis in organisms with large chromosomes are shown in Fig. 1.9.

Single-factor genetic ratios and their explanation

Mendel achieved a great clarification of the rules of heredity by an astute selection of experimental material. The pure-breeding strains of peas that he chose showed sharp differences without confusing intermediate forms. Thus flowers were either purple or white, the ripe seeds were either green or yellow and either round or wrinkled, the plants were either tall or dwarf, and so on. His analysis showed that, for each of these traits, the two contrasting characters were determined by a pair of alternative genetic factors that were present together in the first-generation (F_1) progeny of the appropriate inter-strain cross but segregated away from one another in the formation of the F_1 germ cells. With respect to each difference, the F_1 plants resembled one of the parent strains; the factor inherited from that strain seemed to be *dominant* and the alternative factor from the other parent *recessive* – that is without apparent effect. For example, when pure-breeding purple- and white-

scopical evidence, which we need not detail here, that each chiasma represents a point where homologous chromatids have broken and rejoined cross-wise. The *n* chromosome pairs connected by chiasmata are called *bivalents*, and the chiasmata hold because of the apparently sticky attachment of the chromatids of each chromosome along most of their lengths.

4 At the end of diplotene, the bivalents become greatly contracted as the result of further internal coiling of the chromosomes, following which the nuclear membrane disappears and the bivalents come to lie on the equator of the division spindle. Each bivalent appears to be held in equilibrium on the spindle equator by the equal and opposite pull of its two centromeres to the spindle poles. This stage is metaphase of the first division – *metaphase I*.

5 At *anaphase I* the mutual stickiness of the chromatids is overcome, and the two divided chromosomes (*dyads*) of each bivalent are pulled apart by their undivided centromeres to opposite

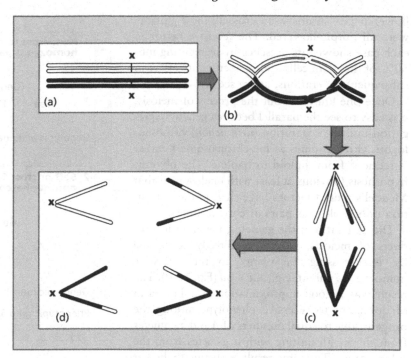

Fig. 1.8 Diagram to show how, as a result of crossing over at chiasmata, each of the four products of meiosis can receive a different combination of maternal and paternal segments of a given chromosome: (a) pachytene; (b) diplotene–metaphase I; (c) anaphase I; (d) anaphase II. Chromosome segments of different parental origins are distinguished as filled versus open outlines. Centromeres are labelled **x**. Note that, through variation in chiasma number and position, and the random involvement of chromatids in successive chiasmata, a large number of different patterns of recombination can be generated from each chromosome pair.

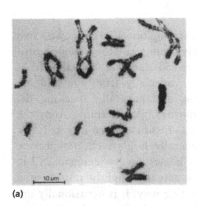

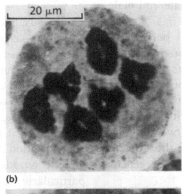

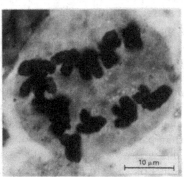

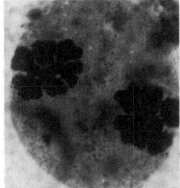

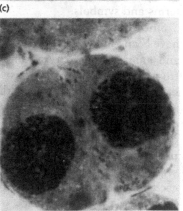

Fig. 1.9 Photographs of some stages of the first division of meiosis. (a) Diplotene in a spermatocyte of the locust, *Locusta migratoria*. Note the chromosomes divided into chromatids and joined in pairs (bivalents) at chiasmata. At each chiasma, two of the four chromatids are exchanged (cf. Fig. 1.6e). (b–e) Metaphase I, early and late anaphase I, and telophase I in pollen mother cells of *Tradescantia paludose* (cf. Fig. 1.6f and g). (a) Courtesy of Dr Gareth Jones.

flowered pea strains were crossed, the F_1 plants were all purple-flowered. The recessive factor in each case showed its presence by segregating into 50% of the germ cells and reasserting its effect in subsequent generations, as we see below.

Once one knows about the process of meiosis, it is easy to see the parallel between genetic segregation and the separation (*disjunction*) of homologous chromosomes as the chromosome number is reduced from diploid to haploid. The obvious hypothesis (obvious at least with hindsight) is that Mendel's paired factors have their physical location in homologous pairs of chromosomes.

The 1:1 ratio in the gametes, the haploid products of meiosis, was more directly confirmed by backcrossing F_1 plants to the recessive homozygous parent, e.g., $Aa \times aa$ (Fig. 1.10). The result was a good approximation to a 1:1 ratio of dominant to recessive phenotype among the progeny, due to equal numbers of A and a gametes from the F_1 all uniting with a gametes from the other parent. The same result is obtained whether the heterozygous F_1 is used as seed parent or pollen parent, thus formally establishing that the 1:1 segregation applies to both female and male gametes.

Before proceeding further it will be well to introduce some essential genetic terminology. Mendel's mutually exclusive factors are called *alleles*, and they are considered to be variants of the same gene. A diploid carrying two identical alleles is said to be a *homozygote* with respect to that particular *gene*, and one with two different alleles is a *heterozygote*. The reader should consult Box 1.2 for a fuller exposition of genetic terms and symbols.

Since Mendel, results formally identical to his have been obtained in experiments on many plants and animals and also on fungi. The information from fungi is actually the most informative, since in many of them, especially those of the Ascomycete class, it is possible to isolate and characterize the products of meiosis in their original groups of four (*tetrads*). The clear prediction of the chromosome theory is that the overall 1:1 segregation should reflect an exact 2:2 ratio in each and every meiotic tetrad. This would seem to be a necessary consequence of the distribution of two divided chromosomes – that is four

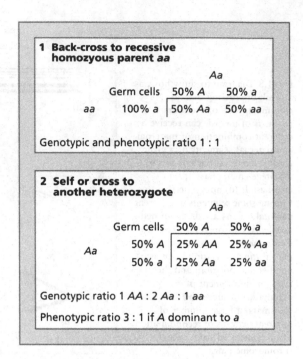

Fig. 1.10 The genetic demonstration of 1:1 segregation of alleles from a heterozygote Aa in germ cell formation.

chromatids, two carrying one alternative and two the other – equally among the four meiotic products. Extensive tetrad data are available from Ascomycete fungi, especially from the budding yeast *S. cerevisiae*, and generally speaking the prediction is confirmed. An example from the fungus *Neurospora crassa* is shown in Fig. 1.11.

Although it generally holds well enough, the generalization that alleles always segregate 2:2 in meiotic tetrads is actually slightly untrue in a particularly interesting way. It is occasionally upset by a low-frequency phenomenon called *gene conversion* which we touch upon in the next chapter. This is of great significance for understanding the mechanism of genetic recombination, a topic beyond the scope of this book.

Independent assortment of different allele-pairs

The second of the two principles established by Mendel (sometimes called his Second Law) was deduced from the results of crossing pea varieties distinguished by two allelic differences. One example was his cross between true-breeding yel-

---

Box 1.2 Genetic terms and symbols

. . . .

Genome. The total genetic material of an organism.

Gene. A genetic element with a specific function and occupying a particular site (*locus*) in the genome. (The nature of the gene will be explored in much more detail in Chapters 2 and 3.)

Allele. A particular form of a gene. Alleles of the same gene are said to be allelic to each other and are given the same basic gene symbol, which may be a single letter, abbreviation or acronym. Different alleles of the same gene are distinguished by capital versus lower-case letters (corresponding to dominant and recessive alleles, respectively) and/or different numbers or superscripts for multiple alleles.

Wild-type allele. The fully functional form of the gene found in the wild or normal organism. In fact, there may be many different wild-type alleles, all fully functional and often difficult to distinguish without molecular analysis.

Mutant allele. A form of a gene observed or presumed to have been derived from the wild-type by mutation, either spontaneous or experimentally induced. Mutant alleles are often, to a greater or lesser extent, functionally defective, in which case they are usually (not always) recessive to the wild type.

Homozygote (adjective *homozygous*). A diploid organism carrying identical alleles of a gene.

Heterozygote (adjective *heterozygous*). A diploid carrying two different alleles of a gene.

(N.B. The last two terms usually refer to a particular gene and not to the whole genome; it is of course possible for an individual to be homozygous for one gene and heterozygous for another.)

Genotype. The genetic consitution of the organism.

Phenotype. The visible characteristics of the organism. The term needs qualification in that phenotypes that are superficially identical are often distinguishable by closer, particularly molecular, analysis.

Dominance/recessivity. The expression, in a heterozygote, of the phenotypic effect of one allele to the exclusion of that of the other. Complete dominance means that the heterozygote is phenotypically indistinguishable from the dominant homozygote (but see the qualification under phenotype above). If there is no dominance the heterozygote is exactly intermediate in phenotype between the two homozygotes. With partial dominance the heterozygote is intermediate but more like one homozygote than the other. *Codominance* is the situation where distinct effects of both alleles can be seen in the heterozygote.

continued on p. 16

Box 1.2 *Continued*

A note on gene and allele symbols

Genes have most often been named after the effects of the mutant allele by means of which they were first identified: e.g., *w*, for white eye in *Drosophila*, *sh* for shrunken endosperm in maize, *gal* for inability to use galactose in yeast. In *Drosophila*, genes have sometimes been given capital or lower-case initials according to whether the originally identified mutant allele was dominant or recessive (e.g., *Abd-A* and *abd-B*, see Chapter 7). In the early days of plant genetics genes were often assigned arbitrary letters, with capital and lower case initial letters to distinguish dominant and recessive alleles; e.g., *A/a*, *R/r* governing pigmentation in maize. Latterly, especially in *Drosophila*, the names have sometimes been based on jokes or literary allusions, e.g., *fushi-tarazu*, *Oskar*.

Wild type alleles are distinguished by a + superscript (e.g., *w*⁺) in *Drosophila*, by a capital initial letter in maize (e.g., *Sh*, *A*, *R*), or in yeast by capitalization of the whole symbol (e.g., *GAL*).

When different genes mutate to give the same or similar phenotypes they are often given the same symbol followed by an identifying number, e.g., *Sh1*, *Sh2* (shrunken endosperm) in maize, *GAL1*, *GAL2*, etc., in yeast.

Mutant alleles are usually distinguished by different superscripts, sometimes descriptive of the phenotype of the homozygote (e.g., *w*ᵃ, *w*ᶜʰ, *w*ᵉ for apricot, cherry and eosin eyes in *Drosophila*) but often just combinations of letters and numbers explicable only by reference to laboratory notebooks. In yeast it is more usual to number mutant alleles sequentially, without use of superscripts, e.g., *gal1-1*, *gal1-2*, etc.

Different conventions are followed for different organisms, but something like the *Drosophila* system is most often adopted. Human gene nomenclature is in a state of some confusion.

---

low, round-seeded and green, wrinkled-seeded stocks. The genotypes of the two parents can be represented, using Mendel's symbols, as *AA BB* and *aa bb*, respectively (*A* for round, *a* for wrinkled, *B* for yellow, *b* for green). It should be noted that the bulk of the pea seed consists of the first seedling leaves (*cotyledons*) of the next generation, which can thus be classified and counted with respect to certain characters while still in the pod. The F_1 seeds, with the doubly-heterozygous genotype *Aa Bb*, were all round and yellow; i.e., *A*

and *B* were fully dominant. These seeds were grown into plants, and backcrossed to the *aa bb* parent. Round yellow (*Aa Bb*), round green (*Aa bb*), wrinkled yellow (*aa Bb*) and wrinkled green (*aa bb*) seeds were formed in a close approximation to equal numbers.

This result, which is obtained irrespective of which parent functions as female and which as male, shows two things. Firstly, both pairs of alleles segregate 1:1 in accordance with Mendel's First Law. Secondly their segregation is indepen-

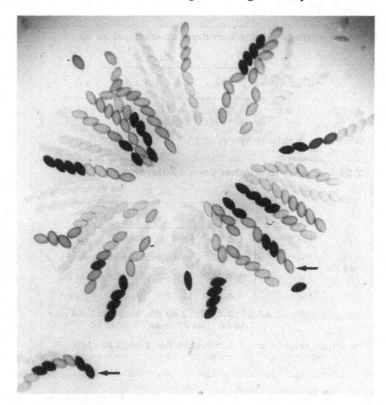

Fig. 1.11 Segregation of alleles governing ascospore colour at meiosis in the ascomycete fungus *Neurospora crassa*. Note that in this species there is a mitotic division following meiosis in the ascus before ascospore formation, so each meiotic product is represented by an adjacent pair of spores. Asci showing segregation at the second division are arrowed (cf. Fig. 1.14). Asci with all spores pale are immature.

dent; the two allelic differences are randomly re-assorted. It does not matter how the double heterozygote is put together; the outcome is the same whether it is obtained from the cross *AA BB* × *aa bb* or from *AA bb* × *aa BB*.

If the double heterozygotes, *Aa Bb*, are crossed among themselves or self-pollinated, the result approximates to a 9:3:3:1 ratio – two independent 3:1 ratios superimposed. This is the expected result if the four equally frequent kinds of germ cell are being paired at random. The reasoning is shown in Fig. 1.12, which also explains various modifications of this standard Mendelian result in cases where the expression of an allele of one gene is dependent upon a second gene, a situation called *epistasis* or *epistasy*.

Mendel's Second Law of independent assort-ment is readily understood in terms of the chromosome theory, provided that the different pairs of alleles are associated with different pairs of homologous chromosomes. There is nothing in the microscopic appearance of meiosis to suggest that the orientation of one bivalent at metaphase I

is anything other than random with respect to the orientation of another bivalent. Indeed, in those rare cases where homologous chromosomes are visibly distinguishable, the randomness of bivalent orientation can be directly demonstrated.

Testing data for fit to theoretical ratios

In the foregoing account of Mendelian genetics, frequent reference has been made to 'approxi-mate' or 'good' fits of data to simple theoretically explicable ratios. A question that arises constantly in simple genetic analysis is whether the fit is good enough for the theory to be plausible. One always expects some deviation from a perfect fit in a progeny of limited size, even if the theory is true. Another way of putting the question is to ask: What would be the probability of getting so large a deviation if one were taking a random sample of the given size from a large population that, as a whole, fitted the ratio exactly? Box 1.3 shows a statistical procedure for calculating this prob-ability. If the answer is less than 0.05, it is a said

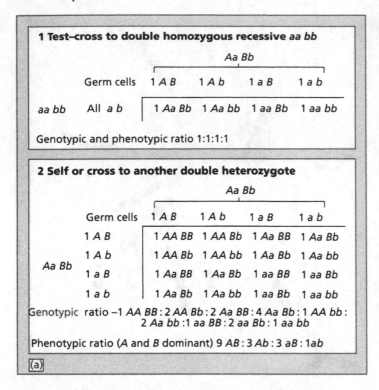

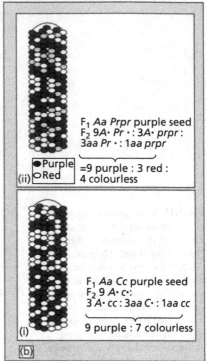

1 Test–cross to double homozygous recessive *aa bb*

		Aa Bb			
	Germ cells	1 *A B*	1 *A b*	1 *a B*	1 *a b*
aa bb	All *a b*	1 *Aa Bb*	1 *Aa bb*	1 *aa Bb*	1 *aa bb*

Genotypic and phenotypic ratio 1:1:1:1

2 Self or cross to another double heterozygote

		Aa Bb			
	Germ cells	1 *A B*	1 *A b*	1 *a B*	1 *a b*
	1 *A B*	1 *AA BB*	1 *AA Bb*	1 *Aa BB*	1 *Aa Bb*
	1 *A b*	1 *AA Bb*	1 *AA bb*	1 *Aa Bb*	1 *Aa bb*
Aa Bb	1 *a B*	1 *Aa BB*	1 *Aa Bb*	1 *aa BB*	1 *aa Bb*
	1 *a b*	1 *Aa Bb*	1 *Aa bb*	1 *aa Bb*	1 *aa bb*

Genotypic ratio –1 *AA BB* : 2 *AA Bb* : 2 *Aa BB* : 4 *Aa Bb* : 1 *AA bb* :
2 *Aa bb* : 1 *aa BB* : 2 *aa Bb* : 1 *aa bb*

Phenotypic ratio (*A* and *B* dominant) 9 *AB* : 3 *Ab* : 3 *aB* : 1 *ab*

(a)

F₁ *Aa Prpr* purple seed
F₂ 9*A· Pr ·* : 3*A· prpr* :
3*aa Pr ·* : 1*aa prpr*

●Purple
○Red   =9 purple : 3 red :
4 colourless

(ii)

F₁ *Aa Cc* purple seed
F₂ 9 *A· c·* :
3 *A· cc* : 3*aa C·* : 1*aa cc*

9 purple : 7 colourless

(i)

(b)

Fig. 1.12 (a) The genetic demonstration of independent segregation at two unlinked loci in a double heterozygote *Aa* and *Bb*. (b) Modifications of two-factor F₂ phenotypic ratios due to interactions of gene effects (epistasis) in maize (*Zea mays*). The cobs shown are the result of self-pollination of two kinds of doubly heterozygous F₁ hybrids: (i) *A/a C/c* and (ii) *A/a Pr/pr*; (i) is homozygous for *Pr* and (ii) is homozygous for *C*. Both dominants *A* and *C* are necessary for anthocyanin (red or purple) endosperm pigment and neither has a visible effect in the absence of the other. *Pr* is necessary for making purple as opposed to red anthocyanin and has no effect in the absence of *A* and *C*. The endosperm mirrors the embryo in its genotype and so one can see the F₂ ratio in the seed on the F₁ cob (see Fig. 1.4). Note that in F₂ kernels, where one dominant allele is present one cannot tell whether the second allele is dominant or recessive. The dots (·) stand for either *A* or *a*, *Pr* or *pr*, *C* or *c* depending on the context.

that the deviation from the theoretical ratio is significant 'at the 5% level'. This does not mean that a valued hypothesis must be discarded (one may just have the unlucky 1-in-20 sample), but at least it suggests the need for more data to either confirm or allay the suspicion about the correctness of the hypothesis.

Linkage groups and the nature of crossovers

As we have seen, the simple principles of Mendelian segregation and independent assortment fit well with the theory that particular inherited differences are associated with particular chromosomes. But, unless the number of genes is limited to the haploid number of chromosomes, independent reassortment cannot always apply. Mendel

studied seven pairs of alleles in pea; luckily for the clarity of his interpretation, they all reassorted independently. But since the haploid chromosome number of pea is only eight he could not, assuming that the chromosome theory is correct, have included many more differences in his study without finding an exception to his Second Law.

In fact, in all plants and animals that have been relatively well-studied genetically, genes have been identified well in excess of the number of chromosomes and have been shown to fall into groups (*linkage groups*) corresponding in number to the haploid chromosome set. Genes that fall into different linkage groups show independent segregation; those in the same group show linkage – that is to say, the combinations of alleles segregating from meiosis tend to be the same as those that

Box 1.3 The chi-squared (χ^2) test for assessing goodness of fit to theoretical ratios

. . . .

If we have a sample of individuals falling into distinct classes, and our hypothesis predicts that the numbers in these classes should conform to a particular ratio, say $1:1$, $3:1$, $9:3:3:1$, etc., we can compare the numbers observed in each class, o_1, o_2, etc., with the 'expected' numbers, e_1, e_2, etc., that would have been a perfect fit to the hypothetical ratio with the given sample size. The statistic χ^2 is calculated by summing the squares of the deviations from expectation divided in each case by the expected number.

$$\chi^2 = (o_1 - e_1)^2/e_1 + (o_2 - e_2)^2/e_2 + \text{etc.}$$

The value of χ^2 has to be judged in relation to the number of degrees of freedom n – that is, the number of opportunites that the data as a whole have for deviating from a perfect fit. Given the sample size, if there are N classes, fixing the numbers in $N - 1$ of them also fixes the Nth class, so the number of degrees of freedom n is always one less than the number of classes.

For a given value of n, each value of χ^2 corresponds to a certain probability (P) of getting so much deviation just by accidents of sampling from a hypothetical large population that fitted the hypothetical ratio. The greater χ^2, the less P; but each additional degree of freedom permits χ^2 to be greater before P becomes unacceptably low.

The following is a sample of χ^2 and P values for different values of n:

	P			
n	0.5	0.1	0.05	0.01
1	0.45	2.70	3.84	6.63
2	1.39	4.60	5.99	9.21
3	2.37	6.25	7.81	11.34

The smaller the value of P, the more confident one becomes that there is something wrong with the hypothesis. For $P < 0.05$, one may say that the deviation is significant at the 5% level (or, for $P < 0.01$, significant at the 1% level), meaning that, if the hypothesis were true, one would only get such an unlucky sample, on average, in one in 20 (or one in 100) trials.

For just two classes, expected in the ratio p:q, the general formula for χ^2 reduces algebraically to:

$$\chi^2 \ (n = 1) = (qa - pb)^2/pq(a + b),$$

continued on p. 20

Box 1.3 *Continued*

where a and b are the observed numbers. For a $1:1$ expectation, χ^2 is just the difference squared divided by the sum.

Examples

1 From crosses between true-breeding plant strains of phenotypes A and a, the F_1 generation is all A and the F_2 consists, from two different crosses, of (i) 82A:33a, (ii) 71A:49a:

(i) is an acceptable fit to a $3:1$ ratio; expected numbers 86.25 and 28.75, deviations ± 4.25, $\chi^2 = 4.25^2/86.55 + 4.25^2/28.75 = 18.06/86.25 + 18.06/28.75 = 0.84$, for $n = 1$, $P = 0.1-0.5$;

(ii) is unacceptable as a $3:1$; χ^2, $n = 1$, calculated in just the same way, comes to 16.9, corresponding to $P \ll 0.01$.

However, (ii) is quite compatible with a $9:7$ ratio, as might arise if the A phenotype were the result of complementary action of two genes (see Fig. 1.12b). Expected numbers 67.5 and 52.5, deviations ± 3.5. χ^2 $(n = 1) = 3.5^2/67.5 + 3.5^2/52.5 = 12.25/67.5 + 12.25/52.5 = 0.41$ $(P = \text{ca. } 0.5)$

2 A test cross of the form $AB/ab \times ab/ab$ gave the following results.

	Phenotypes			
	AB	Ab	aB	ab
Observed numbers	87	32	42	63
Expected for $1:1:1:1$	56	56	56	56
Deviations	31	24	14	7

$$\chi^2\,(n = 3) \quad \frac{31^2 + 24^2 + 14^2 + 7^2}{56} = \frac{961 + 576 + 196 + 49}{56}$$

$= 1782/56 = 31.8$ for 3 df. $(P < 0.01)$. The data are clearly incompatible with $1:1:1:1$.

The overall χ^2 with $n = 3$ can be partitioned into three components, each with $n = 1$:

(i) $1:1$ segregation of $A-a$: observed 119 and 105, $\chi^2 = 14^2/224 = 0.88$ $(P = 0.5-0.1$, insignificant deviation);

(ii) $1:1$ segregation of $B-b$: observed 129 and 95, $\chi^2 = 34^2/224 = 5.16$ $(P = 0.05-0.01$, significant deficiency of b at 5% level);

(iii) $1:1$ ratio of parental to recombinant types: observed 150 and 74. $\chi^2 = 76^2/224 = 5776/224 = 25.78$ $(P \ll 0.01$, highly significant evidence for linkage).

continued

Box 1.3 *Continued*

Thus the overall deviation from $1:1:1:1$ is attributable mainly to linkage between A and B, but also to a deficiency of the b phenotype, presumably due to reduced viability.

Notes

1 The χ^2 method is applicable only to whole-number samples, not to measurements, for which different statistical methods are appropriate (see Chapter 6).

2 The random sampling theory underlying χ^2 depends on the assumption that the numbers in different classes vary symmetrically about their mean values. This becomes increasingly inaccurate as sample sizes become small, since the absolute minimum of zero limits variation at the lower end. As a rule of thumb, the method should not be used if any of the expected numbers is five or less.

entered meiosis. However, even linked genes generally show some frequency of recombination, varying from near-zero to a maximum of 50%, depending on the particular genes involved.

Recombination between linked genes is easily explained as a consequence of the crossing over between chromatids seen at the diplotene stage of meiosis (see Figs 1.6e and 1.9a). It is only necessary to suppose that each gene occupies a distinct position (*locus*) along the length of one of the chromosomes. If so, a difference between alleles segregating at meiosis becomes a *marker* for the gene and the chromosome locus at which it resides.

The hypothesis that recombination between linked markers corresponds to chiasma formation, with two chromatids out of four crossing over at each chiasma, is greatly strengthened by analysis of meiotic tetrads. Extensive data from *S. cerevisiae* and other ascomycete fungi, such as *Neurospora* and *Sordaria* species, demonstrate that tetrads showing recombination between two linked markers are generally *tetratypes* – that is, they are composed of the two parental and the two reciprocally constituted recombinant types (see Box 1.4a and Fig. 1.13). Tetrads with all four spores recombinant occur at a frequency low enough to be explained by double crossing over involving all four chromatids.

A further insight into the nature of crossing over can be obtained from analysis of tetrads segregating at three linked loci, defining two chromosome intervals. The results indicate (Box 1.4a) that, where there is a crossover in each interval, the involvement of a chromatid in one crossover neither increases nor decreases its chance of being involved in the second. Doubles in which the second crossover involves the same two chromatids as the first (so-called two-strand doubles), the two chromatids that were not previously involved (four-strand doubles), and one of the same and one different (three-strand doubles), generally occur in numbers consistent with the $1:1:2$ ratio expected if the choice of chromatids is made at random at each crossover.

The random involvement of chromatids in successive crossovers helps explain the upper limit of 50% for recombination between distant genes. If one chiasma is always formed in a particular interval the result will be 50% recombination between its ends. If a second chiasma is formed, any chromatid that had crossed over at the first will be as likely as not to cross back again; in

Box 1.4 Arguments from tetrad data
. . . .

(a) *Segregation of three linked markers*

Cross *a b c* × + + +

Possible tetrads (written without regard to order within tetrads):

Tetrad class 1   **2**   **3**

a	b	c
a	b	c
+	+	+
+	+	+

a	b	c
a	+	+
+	b	c
+	+	+

a	b	c
a	b	+
+	+	c
+	+	+

Explanation

Non-crossover   1 crossover *a-b*   1 crossover *b-c*

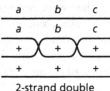

Class 4   **5**   **6**   **7**

a	b	c
a	+	c
+	b	+
+	+	+

a	b	+
a	+	+
+	+	c
+	b	c

a	+	c
a	b	+
+	b	c
+	+	+

a	b	c
a	+	+
+	+	c
+	b	+

2 crossovers
2 chromatids

2 crossovers
4 chromatids

2 crossovers
3 chromatids

2 crossovers
3 chromatids

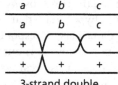

2-strand double   4-strand double   3-strand double   3-strand double

Notes

The map order *a−b−c*, with the implication that tetrads of classes 4–7 represent double crossovers, is deduced from the observation that classes 1, 2 and 3 are much the most frequent.

In the single-crossover classes 2 and 3, only two out of four chromatids are recombined.

Double crossover classes 4–7 are about equally frequent, so all four chromatids participate in crossovers on an equal footing.

continued

Box 1.4 *Continued*

(b) *Tetrad types from two-marker segregations – distinguishing location far apart on same chromosome from location on different chromosomes*

Cross *a b* × + +

Given 2:2 segregation of each marker, there are just three possible kinds of tetrad (leaving aside any order within tetrads):

P: *a b, a b,* + +, + + – parental ditype;
N: *a* +, *a* +, + *b,* + *b* – non-parental ditype;
T: *a b, a* +, + *b,* + + – tetratype.

1 If *a* and *b* are linked:
 no crossing over between *a* and *b* gives a P tetrad;
 a single crossover gives a T tetrad.
 Double crossovers in the *a–b* interval give P, N or T (depending on whether 2, 4 or 3 chromatids are involved – see facing page) in proportions 1:1:2.
 A frequency of crossovers high enough to give as many N as P tetrads (50% recombination) will inevitably give a much larger number of T tetrads. As the mean number of crossovers becomes large, the two *a*s become distributed among the four products at random with respect to the two *b*s. This means a P:N:T ratio of 1:1:4. For short *a–b* distances, N asci are rare; for no *a–b* distance is the ratio T/N less than 4.

2 If *a* and *b* are not linked, the P:N:T ratio depends entirely on the second division segregation frequencies of *a* and *b*. Segregation of *both* at the first division can only give ditype tetrads – P and N equally. If both *a* and *b* are close to their respective centromeres the frequency of T asci will be close to zero. Only if at least one is far from its centromere will T/N become as high as 4.

(c) *Deducing centromere distances from unordered tetrads, as in yeast (see also p. 25)*

If we have two unlinked markers *a* and *b*, and both are segregating from a cross *ab* × ++, the frequency of tetratype tetrads will depend entirely on the two frequencies of second division segregation. Let *p* be the second division segregation frequency of *a* and *q* the second division segregation frequency of *b*.
 There are three ways of getting tetratypes.
1 If *a* segregates at the first division and *b* at the second division.

> *a   b*
> *a* + All tetratypes
> + *b* Frequency $(1 - p)q$
> + +
>   T

2 If *b* segregates at the first division and *a* at the second.

> *a   b*
> + *b* All tetratypes
> *a* + Frequency $p(1 - q)$
> + +
>   T

3 If both segregate at the second division.

> *a b   a b   a* +   *a* +
> + +   + +   + *b*   + *b*   2/4 tetratypes
> *a b   a* +   *a b   a* +   Frequency $pq/2$
> + +   + *b*   + +   + *b*
> P   T   T   N

Total frequency of tetratypes, $T_{ab} = (1 - p)q + p(1 - q) + pq/2 = p + q - 3pq/2$

With a third unlinked marker *c*, with second division segregation frequency *r*, one can determine T_{bc} and T_{ac} and set up the two additional equations:

$$T_{bc} = q + r - 3qr/2 \text{ and } T_{ac} = p + r - 3pr/2$$

Using the three simultaneous equations one can then evaluate *p*, *q* and *r*. The method depends on at least two of the markers showing significantly less than the limiting 67% frequency of second division segregation. Most useful is to find a marker, say *a*, that is very close to its centromere. Then $p = 0$, $q = T_{ab}$ and $q = T_{ac}$.

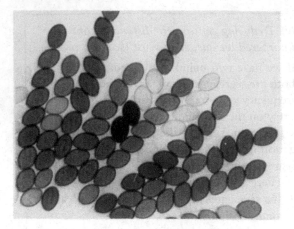

Fig. 1.13 Asci from a cross between two linked pale-ascosopore mutants of the ascomycete fungus *Sordaria brevicollis*, *buff* and *yellow*, not always distinguishable from each other in a black-and-white print. Of nine complete, mature asci, eight are parental ditype, with two *buff* and two *yellow* spore pairs, and one is a tetratype, with one *buff*, one *yellow*, one wild-type (dark brown) and one double-mutant (white) spore pair, resulting from a single crossover between *buff* and *yellow*. There is also one very pale ascus too immature to classify. Note that non-parental ditype asci, with two wild-type and two double mutant spore pairs, would be as common as parental ditypes if the mutants were unlinked, but are rare here because they occur only following four-strand double crossing over between the linked markers (cf. Box 1.4a). Courtesy of Dr D.J. Bond.

other words, any crossover in excess of one will, on average, cancel as many recombinants as it makes.

Map units and map distance

Chiasmata, and the crossovers associated with them, are variable in position from one meiotic cell to another. Sometimes two linked markers will be recombined by a crossover and sometimes not; the probability of their recombination depends on the distance between them. Thus recombination frequencies provide a basis for mapping. The standard map unit is 1% recombination, often called a centimorgan after T.H. Morgan, the pioneer of *Drosophila* genetics.

Recombination frequency is an additive measure of map distance only when it reflects the total crossover frequency in the interval in question. This is the case for short intervals that virtually never contain more than one crossover. For longer intervals, prone to double or multiple crossing over, recombination frequency underestimates crossover frequency because, as explained in Box 1.4, two or more crossovers on average yield the same proportion of recombinant meiotic products as one, namely 50% – two out of four in the meiotic tetrad. The best estimate of the map length of a long interval is obtained by subdividing it by internal markers and summing the estimated lengths of all the subintervals. The more markers, and the smaller the intervals between them, the better the estimate of overall map distance will be.

If it is true, as usually assumed, that all crossovers show as chiasmata at diplotene, map distance is equal to 50 times the mean chiasma count. This was until recently thought to have been confirmed in maize, but recent estimates of total maize map length using DNA markers (see Chapter 5) have come out about 30% higher than the estimate from chiasma frequency (Nilsson *et al.*, 1993); it is not yet clear whether this discrepancy is due to technical error or to some crossovers not resulting in chiasmata.

It should be emphasized that, since crossover density can vary from one chromosome segment to another, map distance cannot be assumed to be proportional to physical distance along the chromosome.

Placing linked genes in sequence

Although recombination analysis can give, at best, only a broad and approximate indication of the physical distances between markers, it can place them in an unequivocal linear order. The standard procedure for determining the order of genes along chromosomes – that is, making a map of the linkage group – is the three-point (or multi-point) *test-cross*. The triple (or multiple) heterozygote is crossed to a tester strain that is homozygous recessive with respect to all the loci concerned.

In the three-point case, there will be eight distinguishable progeny phenotypes, in four reciprocally constituted pairs. One pair consists of the two non-recombinant parental types, two are explicable as due to single crossing over in one or other of the two intervals defined by the three markers, and the remaining pair have to be explained as due to simultaneous crossing over in both intervals. Two of the eight classes are usually conspicuously less frequent than the others; if these are taken to be the double-crossover products the sequence of the three markers is thereby fixed. The reasoning is set out in Box 1.5, together with the calculations of recombination frequency between the markers.

One may ask whether the frequency of double crossovers is consistent with the hypothesis that crossing over in one interval has no effect on the probability of crossing over in the other. If this were the case, the frequency of doubles should approximate to the product of the total frequencies of crossing over in the two intervals separately. In fact, it is common to find fewer doubles than would be expected on this basis, a situation that is described as *interference* between crossovers. The degree of interference is expressed as the *coefficient of coincidence*, which is the observed number of doubles divided by the number that would be expected if crossovers occurred independently of one another. It would be one for no interference and zero for complete interference; it usually takes some intermediate value.

The data shown in Box 1.5 exemplify the point, made in the previous section, that frequencies of recombination between markers are not a strictly additive measure of true map distance. With three markers in the sequence $a-b-c$, the recombination frequency between a and c will be smaller than the sum of the frequencies between a and b and b and c to an extent depending on the frequency of doubles. The more accurate estimate of the map distance $a-c$ is the sum of the estimates of $a-b$ and $b-c$, which themselves may well be underestimates because of undetected double crossovers within these intervals.

Distinguishing between independent assortment and distant linkage

Fifty per cent recombination, the expected result for markers at opposite ends of the same chromosome, is in itself indistinguishable from the free reassortment expected from location on different chromosomes. In the former case, however, it should be possible to connect the distant markers by showing their common linkage to intermediate loci. Where tetrad analysis is possible, as in yeast, a method is available whereby distant linkage can, in some instances, be ruled out. It rests on the argument that, with distantly linked markers, it is impossible to get non-parental ditype asci without a larger number of tetratype asci (see Box 1.4b).

Mapping centromeres using tetrads

One final benefit of tetrad analysis is that it can give information about the behaviour of the centromere and its relationship to linked segregating genes. The centromeres of homologous chromosomes always segregate from each other as undivided wholes at the first division, and split only at the end of second metaphase (see Fig. 1.6). If no chiasma occurs between a marker and the centromere, the alleles will follow the centromere in segregating at the first division – let us say $A-A$ to one pole and $a-a$ to the other. If a chiasma does occur in the gene–centromere interval, the effect will be that the disjoining dyads will carry unlike alleles $(A-a)$ to each pole of the first division spindle. In the latter case the segregation of A from a will be delayed until the centromeres split at anaphase of the second division. In ascomycete fungi like *Neurospora* or *Sordaria* species, meiosis is so arranged as to give ordered tetrads. The second division spindles are aligned more or less end-to-end, so the products of meiosis are formed in a row in the ascus with their positions reflecting the two divisions of meiosis. *First-division segregation* of alleles, call them A and a, will result in two spores carrying A in one half of the ascus and two carrying a in the other half. *Second-division segregation* will result in each half of the ascus containing one A and one a product (Fig. 1.14).

Just as the frequency of recombination between two genes can be used as a measure of the distance between them, so the percentage second-division segregation shown by a single gene can be used as a measure of its distance from the centromere. The centromere can thus be used as a marker and put in sequence with gene markers in a linkage group. It should be noted that, as a measure of distance along the chromosome, percentage second-division segregation has to be divided by two to be equivalent to percentage recombination.

Box 1.5 The three-point test-cross – an example from maize (*Zea mays*)

....

Stock A – homozygous for two recessive mutations, *ligule-less* (*l g*) and *glossy-leaf* (*gl*).

Stock B – homozygous for dominant *Booster* (increased red plant pigment) (*B*).

Cross A × B gives F_1 plant *lg gl b/Lg Gl B*.

Backcross to triple recessive parent *lg gl b/lg gl b*.

Progeny seedlings scored as follows (the phenotypes correspond to the genotypes of the F_1 germ cells):

Seedling phenotype	Number		Simplest interpretation
lg gl b	172 ⎫		Parental combinations clearly exceed all
Lg Gl B	162 ⎬ 334		recombinant classes – three markers
			all linked
lg Gl b	6 ⎫		Uncommon events – these are double
Lg gl B	5 ⎬ 11		crossovers – order must be *lg-gl-B*
lg gl B	56 ⎫	104	Single crossovers between *gl* and *B*
Lg Gl b	48 ⎬		
lg Gl B	51 ⎫	94	Single crossovers between *lg* and *gl*
Lg gl b	43 ⎬		
Total	543		

Recombination

lg/gl = (94 + 11)/543 = 105/543 = 19.3%
gl/B = (104 + 11)/543 = 115/543 = 21.2%
lg/B = (104 + 94)/543 = 198/543 = 36.4%

The best map, with distances in map units, is *lg*–19.3–*gl*–21.2–*B*.

The best estimate of the *lg–B* map distance is 19.3 + 21.2 = 40.5; 36.4 is clearly an underestimate because it excludes the observed double crossovers. The values 19.3 and 21.2 will also be underestimates to the extent that undetectable doubles occur within the *lg–gl* and *gl–B* intervals.

The reason is that it corresponds to chiasma frequency, whereas recombination measures *half* the chiasma frequency since only two chromatids out of four recombine at each chiasma. Like recombination frequency, percentage second-division frequency underestimates distance over long intervals. Markers far from the centromere, with an indefinitely large number of crossovers in the interval, tend to a limiting frequency of 67% (or, more precisely, two-thirds) second-division segregation. There are several possible ways of proving this proposition and readers may like to devise their own.

As was explained in Box 1.4c, information on the positions of centromeres can in principle be obtained from tetrad data even when, as in *Sacccharomyces*, the tetrads are unordered, provided that three or more unlinked genes are segre-

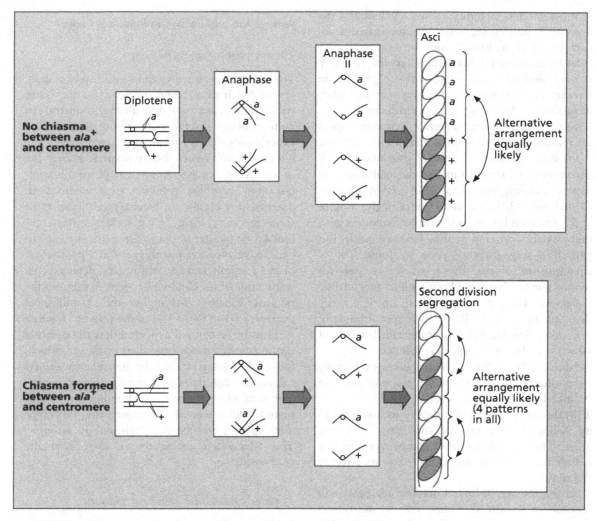

Fig. 1.14 How crossing over between the marker and the centromere leads to segregation at the second division, as seen in an ascomycete fungus such as *Neurospora* or *Sordaria* spp.

gating simultanously and at least one of them is reasonably close to its centromere.

Sex linkage

The great majority of animals and some plants produce approximately equal numbers of males and females. This is the consequence of one of the two sexes, the *heterogametic* sex, carrying an unlike pair of sex chromosomes which are segregated at meiosis to determine two kinds of gametes. The other *homogametic* sex has just one kind of sex chromosome in two copies, and produces only one kind of gamete. There are thus two kinds of fertilization, one leading to the hetero-

gametic and one to the homogametic sex in the next generation. In mammals, insects and certain flowering plants, the heterogametic sex is the male, and the sex chromosomes are called X and Y – XY in the male and XX in the female. In birds and reptiles, on the other hand, the female is the heterogametic sex and the sex chromosomes are called W and Z – WZ in females and WW in males. Here we will deal only with mammals and insects (specifically, the fruit fly *D. melanogaster*).

In male mammals the X and Y chromosomes share a region of homology but differ in substantial segments that are entirely non-homologous. The homologous segments pair at meiosis and generally form chiasmata (cf. Fig. 1.7).

In most species, including mouse and humans, the unique X and Y segments are segregated from each other at the first division of meiosis.

In *Drosophila* the homology between X and Y is very limited. Flies in general are peculiar in having no chiasmata or crossing over in males; nevertheless chromosome pairs, including the X—Y pair, manage to get themselves properly aligned for regular disjunction at the first meiotic division.

In both mammals and *Drosophila*, the Y chromosome carries some of the multiple gene copies encoding ribosomal RNA (rRNA) (see pp. 61–63 and 132) and, more interestingly, in mice and humans it has been shown to harbour a gene that switches development to the male mode. But overall it appears very poor in genes. The X chromosome, on the other hand, is very gene-rich in all animals that have been studied genetically.

Genes on the X chromosome are very easily distinguished genetically by their special mode of sex-linked inheritance, the important features of which may be summarized as follows.

1 Males inherit their X-linked genes exclusively from their mothers, whereas females inherit one X-linked set from each parent.

2 Males cannot be heterozygous with respect to an X-linked gene because they carry only one allele. Thus an allele that would be recessive in a female heterozygote will always show its effect in the male.

3 In the female, X-linked markers segregate and recombine in just the same way as *autosomal* markers, that is to say those located on other chromosomes. In the male they show no recombination because there is nothing for them to recombine with.

4 A few mammalian X-linked genes depart from these rules because they are located on the segment common to the X and Y chromosomes. They are sometimes called *pseudoautosomal* because, like autosomal genes, they are present in two copies in both sexes. Pseudoautosomal markers are sex-linked, but not completely so. They cross over between X and Y chromosomes with a frequency depending on their distance from the junction between the homologous and non-homologous segments.

An example of sex-linked inheritance in *D. melanogaster* is shown in Fig. 1.15.

Segregation and linkage in human genetics

The problem of ascertainment

The rules of gene segregation and linkage apply just as much to our own species as to any other eukaryote but are more difficult to demonstrate in humans because of the impossibility of controlled experiments. The problem is in *ascertaining* the relevant families when there is no prior knowledge of parental genotypes. As an example of this complication, take the common case where both parents of a family are heterozygous carriers of some recessive allele. In all such families combined, one-quarter of the children are expected to exhibit the recessive phenotype. But a proportion of the relevant families will not be detected, because none of the children happens to fall into the minority category. Taking two-child families, for example, 9/16 (3/4 × 3/4) of the relevant families will not be recognized as such unless the parental genotypes are known from their pedigrees, which, for a recessive trait, will be unusual. A similar calculation can be made for each size of family. The number of unascertained families can be estimated from the number in which the phenotype *does* appear, and the data adjusted accordingly. The Mendelian prediction is then generally confirmed.

Linkage assessed by LOD calculations

Human pedigrees showing simultanous segregation of two different markers provide the possibility of testing for linkage and estimating the recombination frequency if linkage appears significant. The situation is, of course, much less tidy than it is in an experimental test-cross. Only some of the groups of brothers and/or sisters (sibships) will be informative, and those that are will seldom comprise more than a few sibs. It may not always be possible to deduce with certainty the relevant parental genotypes, though it may well be possible to assign a probability to each of the possible alternatives.

The best one can do in such a situation is to ask what the probability would be of getting the observed distribution of phenotypes if the recombination frequency took some particular value.

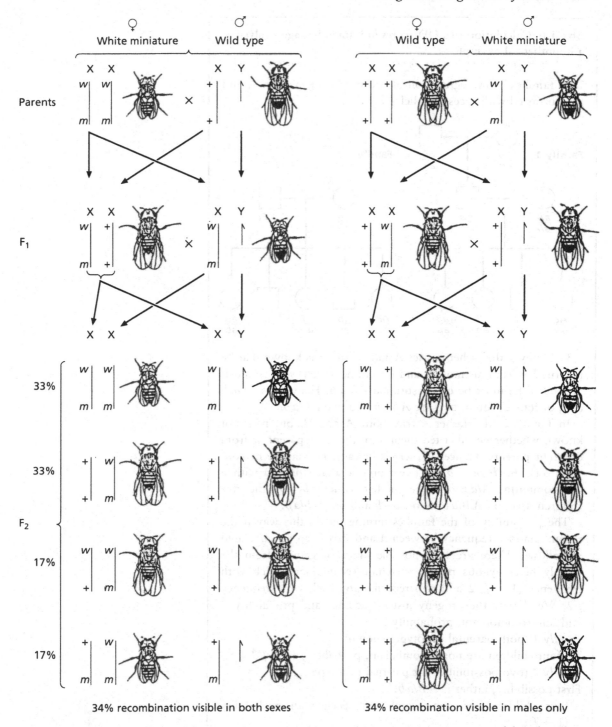

Fig. 1.15 Linked segregation of two X-linked markers, *white-eye* (*w*) and *miniature-wing* (*m*) in *Drosophila melanogaster*. They are separated by about 34 map units (centimorgans). The Y-chromosome does not carry these genes.

**Box 1.6 Calculation of LOD scores in human linkage analysis –
a hypothetical example**

· ·

Two families show segregation of two dominant markers A and
B, contrasted with recessive alleles a and b.

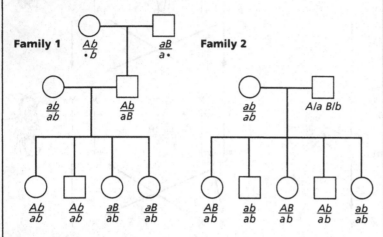

In family 1 the father carries A and B, and it is known that he
inherited A from his mother and B from his father, so if A and B
are linked he must be of constitution $A\,b/a\,B$. His wife is $a\,b/a\,b$
and, of four children, two are $A\,b/a\,b$ and two $a\,B/a\,b$.

In family 2, the father carries both A and B, but it is not
known whether he inherited them from the same parent or from
different parents. Hence, if there is linkage, he may be of constitution $A\,B/a\,b$ or $A\,b/a\,B$ (we represent that situation with a
non-committal $A/a\,B/b$). The mother is $a\,b/a\,b$. Of the five
children, two are $A\,B/a\,b$, two $a\,b/a\,b$ and one $A\,b/a\,b$.

The probability of the families turning out in this way if the
recombination frequency between A and B is f can be calculated
as follows. There are four possible kinds of gamete from the
doubly heterozygous parent, two non-recombinant, each with
frequency $(1 - f)/2$ and two recombinant, each with frequency
$f/2$. We classify the progeny just as recombinant, probability f,
and non-recombinant, probability $(1 - f)$.

Family 1 (both parental genotypes known).
All four children are non-recombinant; probability $(1 - f)^4$.
Family 2 (two possibilities for paternal genotype).
First possibility, father is $A\,B/a\,b$:

  four children are non-recombinant, each with probability
  $(1 - f)$;
  one is recombinant, with probability f;
  probability of the observed sequence* of 4 non-recombinants
  and 1 recombinant $= f(1 - f)^4$.

continued

Box 1.6 *Continued*

Second possibility; father is *A b/a B*:
  four children are recombinant, each with probability *f*;
  one is non-recombinant, with probability $(1 - f)$;
  probability of the observed sequence of 1 non-recombinant and
  4 recombinants = $f^4(1 - f)$.

Giving equal weight to these two possibilities (they are both equally likely if *A* and *B* are in linkage equilibrium, see p. 180):

Average probability of the observed family = $\{f(1 - f)^4 + f^4(1 - f)\}/2$.

Taking both families into account, the overall probability is the product of the two single-family probabilities:

Overall probability = $\{f(1 - f)^8 + f^4(1 - f)^5\}/2$

We can now calculate the overall probability, or odds against, for different values of *f*.

f	Probabilities (each × 10^{-3})	Relative odds	Log odds (LOD)
0.5 (no linkage)	1.95	–	–
0.3	9.33	4.78	0.68
0.2	17.03	8.72	0.94
0.1	21.55	11.04	1.04
0.05	16.58	8.49	0.93
0.02	8.51	4.36	0.64

The odds are fairly strongly in favour of linkage. The recombination frequency is likely to be in the range 5–20% but, with such limited data (only nine tested gametes), cannot be calculated with any precision. As more informative families were taken into account the LOD scores (log odds, relative to no linkage) would usually increase rapidly, with a better-defined maximum pointing to a more precise recombination frequency.

* Note. The calculation is made in this way for the sake of simplicity. Calculating the probabilities of the overall numbers of recombinants and non-recombinants, regardless of their sequence in the family, would have given values 5 times as great for all values of *f*, and so would have made no difference to the relative odds and the final LOD scores.

This is a calculation that a computer can be programmed to perform. One can include in the computation all known pedigrees (there may not be very many of them) that are informative with respect to the combination of markers of interest. The probabilities corresponding to the different pedigrees can be multiplied together to provide an overall probability. The computation is repeated for each of a number of recombination values, spaced, for example, at intervals of 2% over some plausible range – say 5–35%. The overall probability for each recombination frequency, which will usually be very small, is divided by the probability of getting the data if there were no linkage at all, which, if linkage is real, should be even smaller. The resulting ratio, which represents the relative odds against the data given the postulated recombination value, is expressed as the logarithm to base 10 and called the *LOD score*.

A plot of LOD against postulated recombination frequency may show a well-defined maximum that can be taken as the best estimate of the recombination value. The significance of the evidence for linkage depends on the magnitude of the LOD; a value of 2 is about the minimum for acceptability. A simple invented example is shown in Box 1.6.

Linkages involving recessive markers are difficult to establish from human pedigrees because most of the relevant genotypes are concealed. The first linkage maps for humans involved X-linked markers, which always show in male offspring. A few autosomal linkages were established for dominant clinical conditions and blood groups, which are determined by codominant alleles. The recent great development of human linkage maps has depended on differences directly observable in the DNA, at which level problems of dominance do not arise (see Chapters 3 and 5).

Assigning linkage groups to chromosomes

Accepting that linkage groups are equivalent to chromosomes, how can one find out which particular linkage group corresponds to which microscopically identifiable chromosome? Or, more ambitiously, which part of a chromosome harbours a particular gene?

These questions can be addressed by identifying visible changes in the chromosome set and looking for associated changes in patterns of inheritance of genetic markers. Two kinds of chromosomal change are relevant here. Firstly there are structural rearrangement – deletions, segmental inversions, segmental transpositions – resulting from aberrant healing of chromosome breaks which, in experimental genetics, are usually induced by X-rays or other mutagenic treatment. Secondly, there are occasional losses or gains of whole chromosomes (*aneuploidy*) resulting from *non-disjunction* – the occasional accidental passage of sister chromosomes to the same spindle pole rather than to opposite poles at mitosis or meiosis.

Structural chromosome changes are particularly easy to analyse microscopically in *Drosophila* species because of the giant chromosomes that occur in the nuclei of the salivary glands and certain other secretory tissues. These *polytene* (many-stranded) chromosomes have undergone about seven rounds of replication to give an over 100-fold amplification of material without separation of daughter chromatids, which remain fully extended and in close homologous association. Chromosome regions that are more or less compacted in structure show up after staining as darker or lighter bands, and each chromosome is recognizable by its detailed banding pattern. An additional bonus for the investigator is the close somatic pairing of homologous chromosomes, analogous to meiotic pairing, that allows one to see all but the smallest structural changes in one polytene chromosome through the resulting distorted alignment with the homologue (see Fig. 1.17b).

Deletions of chromosome segments are generally lethal when homozygous. When heterozygous, small deletions are often viable, but any recessive mutant allele located in the corresponding segment of the homologous chromosome will show in the phenotype, since there will be no dominant wild-type allele to cover its effect. In *Drosophila*, the end-points of all but the smallest deletions can be defined microscopically, in the polytene chromosomes. If a set of overlapping deletions all uncover the recessive alleles of a

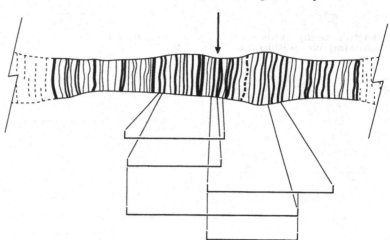

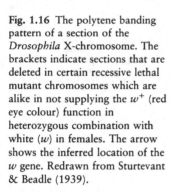

Fig. 1.16 The polytene banding pattern of a section of the *Drosophila* X-chromosome. The brackets indicate sections that are deleted in certain recessive lethal mutant chromosomes which are alike in not supplying the w^+ (red eye colour) function in heterozygous combination with white (w) in females. The arrow shows the inferred location of the w gene. Redrawn from Sturtevant & Beadle (1939).

particular gene, that gene can be assigned to the chromosome segment in which they all overlap, consisting perhaps of only one or a few bands (e.g., Fig. 1.16). In organisms without polytene chromosomes, deletions that are small enough to be viable in heterozygotes and yet large enough to see are much more difficult to obtain.

Other kinds of chromosomal rearrangements – segmental *inversions* and *interchanges* – also provide opportunities for correlating linkage groups with visible chromosomes. Diploids that are heterozygous with respect to the inversion of a chromosome segment may form a visible inversion loop, either at pachytene of meiosis or (in the case of *Drosophila*) in the paired polytene chromosomes (Fig. 1.17b). Since meiotic crossing over within an inversion loop leads to predominantly inviable products with duplications and deletions, inversion heterozygotes show greatly reduced or zero recombination of markers located within the inversion. In *Drosophila*, heterozygous inversions that are confined to one chromosome arm (*paracentric* inversions) are completely viable because of a combination of two circumstances: there is no crossing over in the male (a bizarre property of flies) and the inviable crossover products in the female are, for mechanical reasons, always discarded in the polar bodies (Fig. 1.17a). The viability of *Drosophila* paracentric inversions makes them extremely valuable

for the construction of balancer chromosomes, used for the maintenance of stocks carrying complementary linked recessive lethal mutations (see Fig. 2.2).

Transpositions of segments from one chromosome to another are usually reciprocal, although the smaller of the exchanged segments is sometimes hard to see. In meiosis in an interchange heterozygote, i.e., a diploid with two noninterchanged and two interchanged chromosomes, one usually sees groups of three or four chromosomes joined by chiasmata (*trivalents* and *quadrivalents*) as well as the usual bivalents. Most modes of chromosome disjunction result in inviable meiotic products with one segment duplicated and the other one missing altogether. The only viable products recovered will be those resulting from the segregation, at the first division of meiosis, of the two interchanged chromosomes to one pole and the two structurally normal ones to the other, as can occur through an alternating or 'figure-of-eight' orientation of a quadrivalent at metaphase I (Fig. 1.18). Thus, two normally independent linkage groups become tied together, and can recombine with each other only in so far as crossovers occur between the markers and the segmental exchange points.

Of the various possible types of aneuploid, *trisomics* which have one extra chromosome ($2n + 1$) have been useful for assignments of genes to

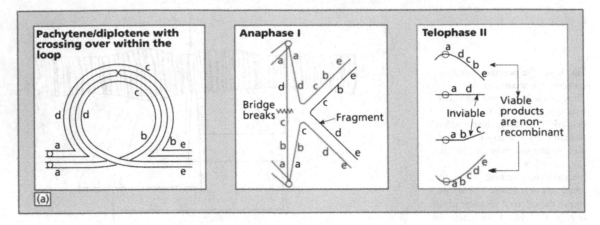

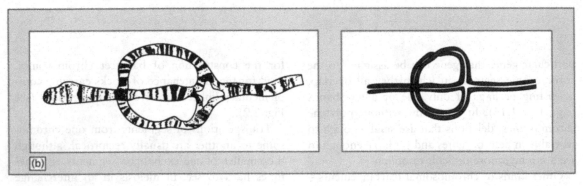

Fig. 1.17 (a) Heterozygosity with respect to an inversion within one chromosome arm (paracentric) supresses recombination within the inversion. Single crossover products are inviable because of bridge-and-fragment formation and consequent generation of chromosome deficiencies. Double crossovers involving the same chromatid are viable, but usually rare. In *Drosophila* such inversions survive because there is no crossing over in the male, and in the female bridge formation affects orientation of the dyads on the second-division spindle, so that the inviable product is discarded in the polar body. (b) A polytene chromosome inversion loop seen in a *Drosophila melanogaster* salivary gland preparation from a female heterozygous for a paracentric X-chromosome inversion. Drawn from the photograph in Sturtevant & Beadle (1939). Diagrammatic interpretation to the right.

chromosomes in various flowering plants. Complete sets of trisomic strains, one for each member of the haploid chromosome set, are available in several species, including tomato and maize. The trisomics are all distinguishable phenotypically. Trisomic heterozygotes show 2:1 or 1:2 ratios for markers on the chromosome present in three copies, easily distinguished from the 1:1 ratios expected for markers on any of the other chromosomes. Trisomic analysis is also a standard method in *Saccharomyces* genetics. In animals, however, trisomics are usually inviable. The only viable human trisomic condition, three copies of

chromosome 21 – Down's syndrome – is associated with some minor physical abnormalities and mental retardation.

Genetic analysis of bacteria and bacteriophage

Three modes of gene transfer in bacteria

Most of the important features of bacterial genetics can be illustrated by reference to the enteric bacterium *Escherichia coli*. This most thoroughly investigated bacterial species is, as we

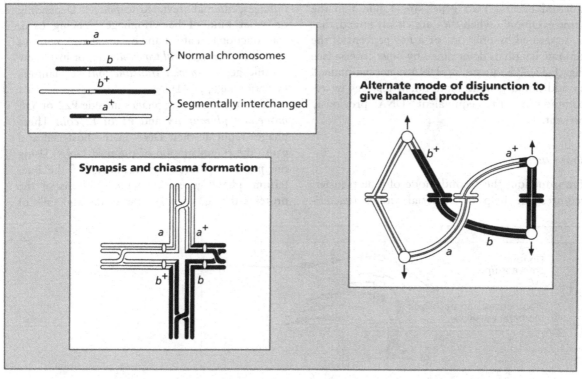

Fig. 1.18 A segmental interchange effectively ties linkage groups together. At metaphase/anaphase I only alternate orientation of centromeres, with the interchanged chromosomes passing to one pole and the non-interchanged chromosomes to the other, give balanced and viable chromosome sets. Hence the markers a/a^+ and b/b^+, which would normally reassort independently, are effectively linked and can recombine only when a crossover occurs between one of them and the interchange breakpoint. A linkage map constructed on the basis of segregation from such a heterozygote would be cross-shaped, mirroring the pattern of chromosome pairing.

see in Chapter 3, also of very great importance as an essential tool in the molecular genetics of eukaryotes.

Bacteria do not have a sexual cycle in the sense of a regular alternation of haploid and diploid phases. What they do have is a variety of means for occasional – generally rare – transfer of *fragments* of their genomes from one cell to another. Such transfer can occur in three different ways.

Transformation

Transformation is the name given to the stable transfer of single genes or small groups of linked genes as free DNA. It was first demonstrated for the mouse pneumonia bacterium, *Diplococcus pneumoniae*, and was the first strong evidence that DNA could carry heritable information. Sub-

sequently, many other bacterial species were shown to be transformable, although some would only assimilate DNA under special conditions.

The standard protocol for transforming *E. coli* involves prior treatment of the cells with calcium ions. Transformation can be obtained with respect to any gene for which selection can be made. For example, a mutant strain of *E. coli* that is unable to grow without a nutritional supplement because of a gene mutation can be treated with DNA isolated from the wild-type organism and transformed cells will grow to form colonies on unsupplemented medium. Transformation frequency may be of the order of one in 100 or 1000 cells, depending on the DNA concentration.

Double transformation for two traits at the same time usually only occurs with the very low frequency expected from two separate, coincidental events. Two genes are acquired in the same event only when they are closely linked, that is, separated by only one or a few per cent of the genome length as determined by other criteria (see Fig. 1.22). The incoming DNA is not maintained as additional genetic material; it replaces by recombination the equivalent DNA previously present.

Transduction

Transduction, the second mode of gene transfer, requires the help of bacterial viruses (*bacteri-*

ophages, often abbreviated to *phage*). Depending on what kind of bacteriophage is being used, transduction can affect any bacterial gene without discrimination (*general transduction*) or just a few specific genes (*special transduction*; see lambda bacteriophage, p. 41).

General transducing phages include P22 of *Salmonella typhimurium* and P1 of *E. coli*. These phages replicate their DNA in the host cell and cause the fragmentation of the host DNA. Using the protein synthesizing machinery of the bacterium, phage genes direct the synthesis of the proteins that go to make the heads and tails of

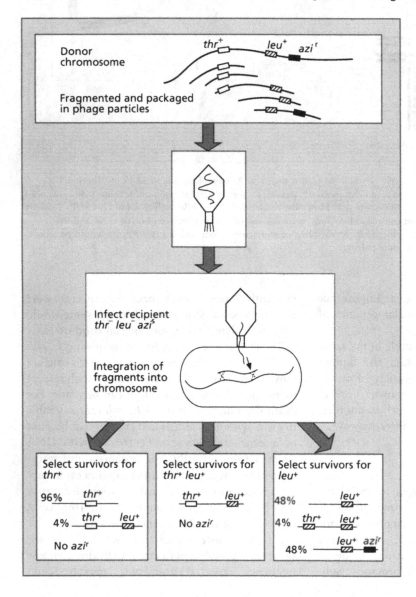

Fig. 1.19 Establishment of the sequence of closely linked markers in the bacterium *Escherichia coli* by cotransduction with bacteriophage P₁. Cells of genotype *leu⁻ thr⁻ azi*ˢ (requiring leucine and threonine and sensitive to azide) were infected with phage that had been grown on a donor strain of genotype *leu⁺ thr⁺ azi*ʳ (not requiring either amino acid and azide-resistant) and plated on medium selective for either *leu⁺* or *thr⁺* or both, and the colonies that grew up were tested for the unselected markers. The marker order is *thr–leu–azi*, and the length of chromosome between *thr* and *leu* (about 2% of the complete genome, cf. Fig. 1.22) is close to the maximum that can be accommodated within the phage head.

the tadpole-shaped phage particles, which are released when the infected cell undergoes dissolution (*lysis*). The phage heads contain the phage DNA and the tails are the organs of infection, sticking to the bacterial cell wall and injecting the phage DNA.

Transduction comes about because, with a certain frequency, fragments of bacterial DNA get packaged in phage heads 'by mistake', and so some of the phage particles released from lysed cells transmit not virus but bacterial DNA to newly infected cells (Fig. 1.19).

In effect, general transduction is simply transformation by other means. Because of the limited capacity of the phage head, only small linked groups of bacterial genes can be transferred together. This feature of transduction makes it suitable for small-scale genetic mapping (Fig. 1.19 shows an example). Transduced genes, like those acquired by transformation, persist through cycles of cell division only if integrated into the recipient genome by homologous recombination.

Conjugation

Conjugation, that is transient cell-to-cell contact and communication, provides the third means of gene transfer in bacteria. If certain strains of *E. coli* with different genetic deficiencies (e.g., nutritional requirements) are mixed and allowed to make contact, wild-type recombinants can be selected from the mixture. The ability of *E. coli* cells to transfer and recombine genetic markers depends on the presence in at least one parental strain of a fertility factor F. In a 'cross' between a strain carrying (F$^+$) and one lacking it (F$^-$), bacterial gene markers are transferred, generally with a rather low efficiency, from the F$^+$ to the F$^-$ partner. At the same time, the F$^-$ factor itself is transferred with very high efficiency, so that virtually all of the initially F$^-$ cells become F$^+$. It turns out that F is a *transmissible plasmid*, a closed loop molecule of DNA that determines a mechanism for its own cell-to-cell transfer. This mechanism includes the synthesis on the surface of the F$^+$ cell of a special filament, a *pilus*, which can attach to another cell, and an enzyme complex that replicates the F$^-$ DNA, starting at a fixed point and generating a single strand that enters the other cell

and becomes double stranded by new synthesis after entry (see Fig. 1.21a).

Time-of-entry mapping in *E. coli*

Escherichia coli F$^+$ strains produce occasional subclones that yield greatly increased numbers of recombinants in crosses to F$^-$ strains. Each high-frequency recombination (Hfr) clone transfers a particular subset of genetic markers in a particular time sequence.

A simple example is shown in Fig. 1.20. In this experiment, the F$^-$ strain is a quadruple mutant; it is resistant to the antibiotic streptomycin (*str*^{r}), it

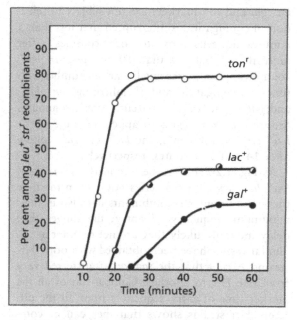

Fig. 1.20 An example of time-of-entry mapping in *Escherichia coli*. Hfr *leu*$^+$ *lac*$^+$ *gal*$^+$ *ton*^{r} *str*^{s} (able to synthesize leucine and to use lactose or galactose as carbon source, phage T1-resistant, streptomycin-sensitive) *tr*^{s} and F$^-$ *leu*$^-$ *lac*$^-$ *gal*$^-$ *ton*^{s} *str*^{r} strains were mixed and, after different times, conjugational contacts were broken by mechanical agitation and the cells were plated on medium lacking leucine and containing streptomycin. This selected recombinant cells with *leu*$^+$ from the Hfr and *str*^{r} from the F$^-$ parent. Colonies that grew up were tested for presence of the other Hfr markers: T1-resistance (*ton*^{r}) and ability to ferment galactose (*gal*$^+$) and lactose (*lac*$^+$). No colonies at all appeared after less than 8-minutes uninterrupted contact. Note that *leu* happened to be the closest of the four markers to the point of initiation of transfer in this particular Hfr strain (HfrH) (see Fig. 1.22).

requires leucine as a nutritional supplement (*leu⁻*), and is unable to utilize the sugars lactose and galactose (*lac⁻*, *gal⁻*). It is wild-type in its sensitivity to bacteriophage T1 (*ton*ˢ). The Hfr strain is wild-type in its sensitivity to streptomycin (*str*ʳ) and in its ability to grow without leucine (*leu⁺*) and to utilize both sugars (*lac⁺*, *gal⁺*), and it carries a mutation conferring resistance to T1 (*ton*ʳ). Suspensions of cells of the two strains are mixed and allowed to conjugate undisturbed for different periods of time, after which they are mechanically agitated to separate conjugating cells, diluted and plated on growth medium containing streptomycin but without leucine. These conditions select for recombinants that inherit *str*ʳ from the F⁻ and *leu⁺* from the Hfr parent – the *lac*, *gal* and *ton* markers are not subject to selection. If conjugation is interrupted after less than 5 minutes no cells grow to form colonies. After more than 5 but less than 10 minutes, *str*ʳ *leu⁺* colonies appear, increasing linearly in number with time of conjugation and all inheriting *ton*ˢ, *lac⁻* and *gal⁻* from the F⁻ parent. After 10 minutes the *ton*ʳ marker begins to appear in some of the *leu⁺* *str*ʳ recombinants, and *lac⁺* and *gal⁺* follow after 14 and 21 minutes, respectively.

As Fig. 1.20 shows, the Hfr markers *leu⁺ ton*ʳ *leu⁺ lac⁺ gal⁺* form a series not only in the time they take to enter recombinants but also in their maximum frequency of entry; the longer they delay the more likely they are not to enter at all. Similar results have been obtained with other Hfr strains, except that the markers transferred vary from one Hfr to another. Comparison of all the marker time-of-entry sequences from all the different Hfr strains shows that they can be combined to form a circular time map measuring about 100 minutes in circumference. Each Hfr strain transfers markers from a sector of the map, starting at a fixed point and proceeding either clockwise or anticlockwise. Transfer can be broken off at any time, and so the further a marker is from the leading point the less likely it is to be transferred.

What is the role of the F-factor in Hfr marker transfer? Two findings help to answer this question. Firstly, whereas the F-factor in an ordinary F⁺ strain is highly infectious by cell contact, in an Hfr strain it is hardly transferred to F⁻ cells at all.

However, if selection is made for the very rare recombinants receiving an extreme 'tail-end' marker after about 100 minutes of conjugation, the F-factor is very often found to have been transferred as well. These observations are strong evidence for the hypothesis that the Hfr state arises from the integration of the F-plasmid into the chromosome at some point – different in different Hfr strains. Then the same mechanism as is normally responsible for the transmission of F from cell to cell (Fig. 1.21a), will transfer a part of F followed by the adjacent chromosome segment and eventually, in the rare event of the entire *E. coli* genome being transferred, by the remainder of F. Figure 1.21b illustrates the hypothesis, which is now generally accepted.

Time-of-entry mapping defines the overall form of the *E. coli* genome (Fig. 1.22). It predicts that it should consist of a single closed-loop ('circular') DNA molecule, a prediction that has been verified by the gentle release from cells of radioactively labelled DNA, followed by autoradiography. It is usual to call this the *E. coli* 'chromosome', even though it does not conform to the original microscopists' definition (see p. 1). The accuracy of the time map over short distances is restricted by the limited precision of the time experiments. For estimating distances and sequences over shorter intervals – fractions of a minute on the time map – transduction and, latterly, DNA sequencing have been more useful.

Analogy with sexual recombination and segregation

Although bacteria do not have sex in the strict sense of alternation of haploid and fully diploid phases, there is a clear parallel between the systems just considered and meiotic recombination.

The *E. coli* chromosome has been shown by autoradiography of the extended DNA to have only a single origin of replication, and so most of the segments transferred either from an Hfr strain or by transduction or transformation, will be unable to replicate. However, *E. coli* cells (and probably bacterial cells generally) are very efficient at recombining homologous chromosomes or chromosome fragments. Such recombination is analogous to meiotic crossing over and gene conversion in eukaryotes. A fragment of chromosome introduced into an F⁻ from an Hfr cell usually undergoes a number of crossovers with the previously resident chromosome, and this results in integration into the chromosome of some, though not usually all, of the Hfr markers (Fig. 1.21c). The displaced F⁻ markers are lost.

To summarize, sexual reproduction consists of the cycle haploid → karyogamy → diploid → meiosis → haploid. *Escherichia coli* does not have anything so regular, but it can from time to time undergo the sequence haploid → transfer of DNA fragment → partial diploid → recombination → reduction to haploid. This has much the same

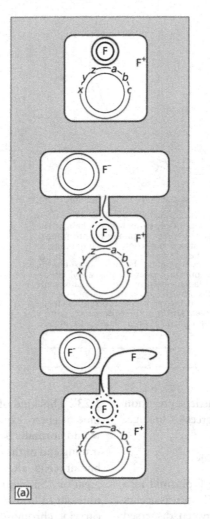
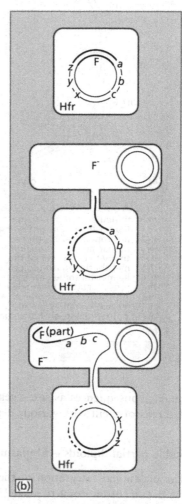

Fig. 1.21 (a) Transfer, by replication, of the F-plasmid from one *Escherichia coli* cell to another. The closed-loop DNA of F is nicked in one strand, which is then displaced by new synthesis (broken line) and transferred through a conjugational bridge to the F⁻ cell. The complete single-strand F copy is made double-stranded and recircularized after entry. The chromosome and the F plasmid (not drawn to scale) are distinguished as thin and thick lines.

(b) How, in an Hfr strain, in which F is integrated into the chromosome between markers *z* and *a* in the sequence *x–y–z–a–b–c*, the F-transfer mechanism results in transfer of a chromosome segment starting with marker *a* and extending for variable distances but very rarely seldom reaching as far as *z*.

(c) Examples of integration of markers from the transferred chromosome segment by crossing over. Donor and recipient alleles are distinguished by capital versus lower-case letters. Crossovers are usually numerous, so even if markers are transferred together they tend to be integrated independently when separated by more than about 5% of the genome.

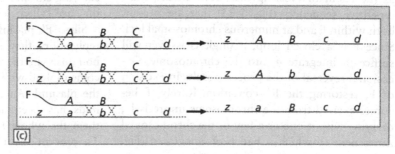

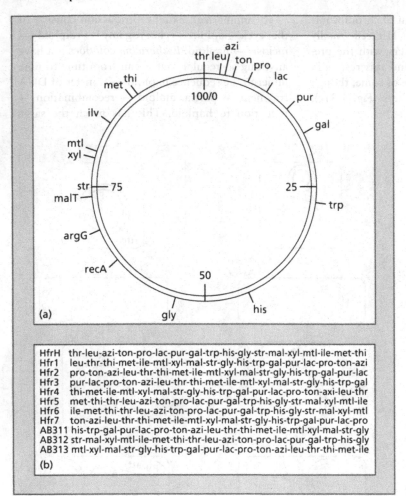

HfrH thr-leu-azi-ton-pro-lac-pur-gal-trp-his-gly-str-mal-xyl-mtl-ile-met-thi
Hfr1 leu-thr-thi-met-ile-mtl-xyl-mal-str-gly-his-trp-gal-pur-lac-pro-ton-azi
Hfr2 pro-ton-azi-leu-thr-thi-met-ile-mtl-xyl-mal-str-gly-his-trp-gal-pur-lac
Hfr3 pur-lac-pro-ton-azi-leu-thr-thi-met-ile-mtl-xyl-mal-str-gly-his-trp-gal
Hfr4 thi-met-ile-mtl-xyl-mal-str-gly-his-trp-gal-pur-lac-pro-ton-axi-leu-thr
Hfr5 met-thi-thr-leu-azi-ton-pro-lac-pur-gal-trp-his-gly-str-mal-xyl-mtl-ile
Hfr6 ile-met-thi-thr-leu-azi-ton-pro-lac-pur-gal-trp-his-gly-str-mal-xyl-mtl
Hfr7 ton-azi-leu-thr-thi-met-ile-mtl-xyl-mal-str-gly-his-trp-gal-pur-lac-pro
AB311 his-trp-gal-pur-lac-pro-ton-azi-leu-thr-thi-met-ile-mtl-xyl-mal-str-gly
AB312 str-mal-xyl-mtl-ile-met-thi-thr-leu-azi-ton-pro-lac-pur-gal-trp-his-gly
AB313 mtl-xyl-mal-str-gly-his-trp-gal-pur-lac-pro-ton-azi-leu-thr-thi-met-ile

(b)

Fig. 1.22 (a) Overall time map of the *Escherichia coli* genome, made by piecing together sequences from different Hfr strains. The map is calibrated in minutes. Only a small selection of the large number of mapped markers is shown. (b) Sequences of markers injected by different Hfrs and their orders of transfer. Piecing these sequences together generated the map shown in (a).

genetic consequences as sex – genetic segregation and reassortment with various degrees of linkage.

Stable partial diploids – F' plasmids

The occasional integration of the F plasmid into the *E. coli* chromosome to make Hfr strains occurs as the result of crossing over between dispersed repetitive sequences (Fig. 1.23) that are present both within F and at numerous chromosomal loci. Since F is a closed loop, a single crossover will suffice to integrate it into the chromosome. Occasional reversal of the process results in excision of F, restoring the F⁺ condition. Rarely, F becomes excised from the chromosome imprecisely, carrying with it one or a few of the chromosomal genes next to which it had been integrated (Fig.

1.23). This kind of event is sometimes detected in time-of-entry experiments when a marker that would normally be transferred, if at all, only at the trailing end of the chromosome, appears in recombinant cells after only a few minutes of conjugation. Such queue-jumping markers generally turn out to be associated with F plasmids that are carrying chromosomal material, or F' plasmids as they are called.

Since F' plasmids have their own origins of replication, they can maintain themselves autonomously in the recipient cells, which thus become diploid with respect to any genes present in the plasmid. An extensive collection of F' plasmids has been accumulated, carrying between them the whole *E. coli* genome in small pieces. To prevent the F's from recombining with the

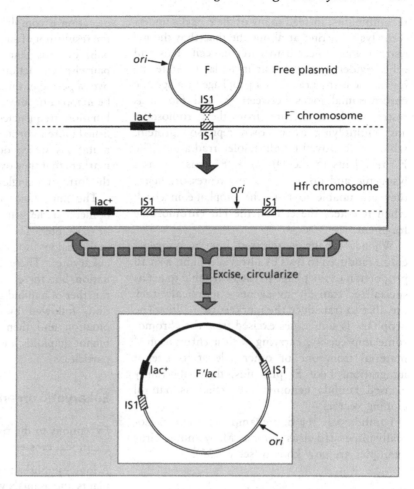

Fig. 1.23 Origin of an Hfr strain, by integration of the F-plasmid close to the *Escherichia coli lac* locus, and the subsequent origin of a F'-lac plasmid by excision and circularization of the chromosome segment, including F, between the arrows. F' plasmids have the F-plasmid origin of replication (*ori*), and so can replicate independently of the chromosome, and they also carry chromosomal markers, *lac* in this case. Integration of F occurs by recombination between dispersed repetitive sequences (e.g., IS1) present on both the chromosome and the plasmid, and it is likely that aberrant excision to form F' derivatives involves similar rare recombination events.

homologous sequences in the chromosome, they are maintained in a mutant *E. coli* strain that is defective in recombination.

Except that they are the products of nature rather than experimental manipulation, F' plasmids are closely analogous to the artificial cloning constructions described in Chapter 3.

Bacteriophage lambda – a virus in the chromosome

We shall not, in this book, be giving more than cursory attention to the genetic systems of viruses, either of bacteria or of eukaryotes. However, two bacteriophages must be at least briefly considered because of their importance in the development of genetic concepts.

Temperate phages, of which bacteriophage

lambda (λ) is the paradigm case, are so called because they can persist in the host cell in a latent state as an alternative to lysing it. In its latent form λ is described as lysogenic, because certain damaging treatments, notably ultraviolet irradiation, will trigger its lytic proliferation.

Important insights into the nature of the lysogenic state were obtained from Hfr × F⁻ crosses in which one parent was lysogenic for λ and the other not. In the cross Hfr × F⁻(λ), with the parents distinguished by markers in the *leu–gal* sector of the genome (see Fig. 1.22), the lysogenic state was inherited just as if it were a marker on the chromosome closely linked to *gal*. More precise mapping by P1 transduction placed it between *gal* and another marker *bio* (nutritional requirement for the vitamin biotin). In the reciprocal cross, Hfr(λ) × F⁻, there were no lysogenic

recombinants; instead many of the F⁻ cells under-
went lysis starting at about the time that the *gal*
marker was expected to enter the cell. This and
other evidence shows that in its latent state the
λ genome is integrated as a prophage at a specific
chromosomal locus between *gal* and *bio*. It is
restrained from excising from the chromosome
and multiplying by its own repressor protein,
which is destroyed by ultraviolet irradiation. The
F⁻ recipients in the Hfr(λ) × F⁻ cross are not
lysogenic and do not have any repressor; hence
they are unable to hold the prophage in check
when it enters the cell on the Hfr chromosome
fragment.

When λ phage is released from a lysogenic
culture induced to lysis by ultraviolet light, a small
proportion of the phage particles are able to act as
specialized transducing agents – specifically they
are able to transduce the markers *gal* or *bio*. The
prophage is sometimes excised from the chromosome imprecisely, carrying with it chromosomal
material from one or other side of its site of
integration. Like F′ plasmids, these aberrantly
excised lambda genomes are acting as natural
cloning vectors.

Lambda was the first example of a chromosomally integrated virus genome. Many animal virus
examples are now known (see p. 135).

Bacteriophage T4 – a model recombinational system

Unlike lambda, the *E. coli* bacteriophage T4 is
exclusively virulent with no lysogenic option. A
great many different mutant derivatives of T4
have been isolated, differing from the wild type in
the *E. coli* strains that they are able to lyse and/or
in the rapidity of lysis, expressed in different sizes
of circular clearings (*plaques*) made in a lawn of
bacterial cells following infection by single phage
particles. If *E. coli* cells are infected simultaneously by two different T4 mutants they release
on lysis not only the two original mutant types but
also wild-type and double-mutant recombinants.
During their joint replication in the host cell,
different T4 genomes are evidently able to recombine with one another.

Frequencies of recombination vary widely, the
maximum being in the region of 40%. Their inter-

pretation is complicated by the fact that they are
the results not of single controlled crosses between
pairs of virus genomes but of successive random
pairwise interactions within a mixed population
over a period of time. Nevertheless, mutations can
be mapped to some extent on the basis of recombination frequencies. Map order can be established more rigorously through the use of deletion
mutations which do not recombine at all with
markers that they overlap (cf. p. 56). The map has
the form of a single closed loop.

The importance of T4 genetics in the development of the gene concept is considered in the
next chapter. The point to note here is the formal analogy to the segregational genetics of
eukaryotes. There is no diploid–haploid alternation but there is, in a T4 'cross', a bringing
together of haploid genomes through mixed infection, followed by an opportunity for recombination and then segregation of pure recombinant haploids in the form of progeny virus
particles.

Eukaryotic organelle genetics

Exceptions to the rule of equal results from reciprocal crosses

A long-established rule in the classical genetics of
plants and animals was that, with the understand-
able exception of sex-linkage, male and female
parents contributed equally to the genotype of the
next generation. This generalization was supported by the results of many crosses between
distinct pure-breeding strains of plants and
animals. Leaving aside animal traits known to be
X-linked, the outcome of such a cross was nearly
always the same, irrespective of which strain was
used as female and which as male. This general
result was easily rationalized on the basis that it
was the gamete nucleus that was important for
hereditary transmission and that male and female
gametes, though usually very different in cytoplasmic mass, contributed nuclei equally.

Chloroplast variants

In fact, exceptions to the rule of equal results from
reciprocal crosses have long been known in plants.

Most examples involve maternal, or predominantly maternal, transmission of defects in the structure or chlorophyll content of the chloroplasts. In some cases a cross between normal and chlorophyll-deficient strains gives progeny that conform entirely to the type of the maternal parent, whether it is fully green or yellow. Less frequently, the progeny may appear mosaic, with most of their tissue resembling the maternal parent but with sectors of paternal type.

Such results were assimilated into the mainstream of modern genetics following the recognition of DNA as the repository of genetic information and the discovery that a fraction of plant cell DNA is contained in the chloroplast. Plant embryos usually acquire their chloroplasts, along with their DNA, from the egg, though in some plants (such as *Pelargonium* spp.) there is some transmission through the pollen tube as well.

Chloroplast-based markers as a group can be distinguished from chromosomal ones by their generally maternal transmission, but there is (at least in flowering plants) no natural system of recombination that can be used to map them further. Nearly everything we know about the chloroplast genome, and the genes that it contains, comes from cloning and sequencing of its DNA. We return to the molecular analysis in Chapter 5.

Mitochondrial variants

Mitochondria, the other type of cell organelle to contain DNA, are present in all normal eukaryotic cells. Essentially, they are the organs of respiration, coupling the oxidation of respiratory substrates to the generation of chemical energy.

Saccharomyces cerevisiae, common yeast, is unusual among eukaryotes in being able to live by alcoholic fermentation in the absence of oxygen and so dispense with mitochondrial function. Hence it has been possible to collect a very large number of respiration-deficient mutants with defective mitochondria, and most of these have their genetic basis in the mitochondria themselves. Numerous mutants that are resistant to antibiotics that inhibit respiration also turn out to be mitochondrial.

There is, in yeast, no distinction between male and female gametes; the fusing haploid cells are of different mating types but similar in size, and both contribute mitochondria to the zygote. A diploid formed by sexual fusion of haploids differing in their mitochondria will initially contain a mixture of mitochondrial types. Remarkably, during the first few diploid budding cycles, the mitochondrial lineages become segregated, so that each cell comes to contain a single type of mitochondrion. When the parental haploids differ with respect to more than one mitochondrial character, the diploid segregants tend to include not only the two parental types but also all possible recombinants. From the frequencies of the different recombinant classes it is possible, in principle, to construct a linkage map showing the relative positions of the genetic markers. In practice, the very high frequency of recombination, probably due to its extending over a period of time instead of being limited to one precise moment as in meiosis, means that linkage is demonstrable only for relatively closely placed markers. The linkage map (which turns out to be circular, corresponding to the closed-loop form of the mitochondrial DNA molecule) was completed by deletion mapping, the principle of which is described in Chapter 2.

The mechanisms underlying yeast mitochondrial recombination and segregation are not well understood. The important point to be made, however, is that they operate, not specifically at meiosis, but in the diploid phase immediately following karyogamy. Thus mitochondrial markers differ sharply in their mode of transmission from chromosomal markers.

Apart from *Saccharomyces*, some of the best examples of mitochondrially-inherited traits are in humans. Several kinds of degenerative neuromuscular disease (myopathies and encephalomyopathies) have been shown to be due to deletions in the mitochondrial DNA. These conditions show strictly maternal inheritance, consistent with the transmission of mitochondria exclusively through the egg (Fig. 1.24). They are not immediately lethal because patients' cells contain some normal mitochondria together with the defective ones, a condition described as *heteroplasmic*. Unlike the situation in *Saccharomyces*, the different mitochondrial lineages do not segregate out during cell division, and a mixed population of mitochondria

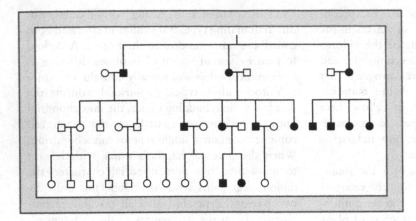

Fig. 1.24 A human pedigree showing the inheritance of a neuromuscular disease (filled symbols) attributed to a mutation in the mitochondrial genome. Females shown as circles and males as squares. Note that affected fathers do not transmit the disease, but affected mothers pass it on to all their children. Data from Giles *et al.* (1980).

can be maintained and transmitted maternally from generation to generation.

Summary and perspective

This chapter has been concerned with the use of natural genetic systems to analyse the modes of organization and transmission of the genes whose variation underlies inherited differences.

The basic method of classical genetics is to bring alternative sets of genes together by sexual or quasi-sexual crosses, and then to observe the patterns of reassortment or linkage that they display when they segregate away from one another again during the natural life histories of organisms.

Such observations, often aided by microscopy, show that genes are located in linear chromosomes in eukaryotic nuclei and in analogous closed-loop structures in bacteria. Except in yeast, the closed-loop subgenomes of eukaryotic organelles required DNA analysis for their demonstration (see Chapter 3).

The mapping of genes along the chromosomes is made possible by the universal phenomenon of recombination – the propensity of chromosomes (or chromosome fragments) to undergo homologous exchanges at frequencies related to chromosome length.

The establishment of the gene as an extended segment of a chromosome, subdivisible by mutation and recombination, is the subject of our next chapter. The substantiation of the concept of genes as segments of DNA, first strongly indicated by the natural process of bacterial transformation,

required the development of a new genetics based on DNA technology, to be described in Chapter 3.

Further reading

Mendelism and chromosome theory

Sturtevant, A.H. & Beadle, G.W. (1939) *An Introduction to Genetics*. W.B. Saunders, Philadelphia. Reprinted 1962, Dover Publications, New York. (Never surpassed as a clear and authoritative account of basic chromosome theory.)

Bacterial genetics

Hayes, W. (1968) *The Genetics of Bacteria and their Viruses*, 2nd edn. Blackwell Scientific Publications, Oxford. (The classical book – never adequately replaced.)

Fungal genetics

Fincham, J.R.S., Day, P.R. & Radford, A. (1979) *Fungal Genetics*, 4th edn. Blackwell Scientific Publications, Oxford. (The detailed background to the fungal systems.)

Organelle genetics

Gillham, N.W. (1978) *Organellar Heredity*. Raven Press, New York.

Chromosome structure

van Holde, K.E. (1988) *Chromatin Structure and Function*. Springer-Verlag, New York. (Chapter 7, pp. 289–343, is highly relevant to this chapter.)

References

Boy de la Tour, E. & Laemmli, U.K (1988) The metaphase scaffold is helically folded; sister chromatids have predominantly opposite helical handedness. *Cell*, 55, 937–44.

Giles, R.E., Blanc, H., Cann, H.M. & Wallace, D.C. (1980) Maternal inheritance of human mitochondrial DNA. *Proc Natl Acad Sci USA*, 77, 6715–19.

Marsden, M.P.F. & Laemmli, U.K. (1979) Metaphase chromosome structure: evidence for a radial loop model. *Cell*, 17, 849–58.

Nilsson, N.O., Sall, T. & Bengtsson, B.O. (1993) Chiasma and recombination data in plants: are they compatible? *Trends Genet*, 9, 344–8.

Srb, A.M., Owen, R.D. & Edgar, R.S. (1965) *General Genetics*, 2nd edn. W.H. Freeman, San Francisco.

Widom, J. & Klug, A. (1985) Structure of the 300 Å chromatin filament: X-ray diffraction from oriented samples. *Cell*, 43, 207–13.

2

....

FROM MUTATIONS TO GENES

Defining the gene by mutation and complementation

What is it that the markers mark?

In the last chapter we saw how, through the use of natural genetic systems, inherited variation can be analysed in terms of unit differences that can be mapped to linkage groups, corresponding to chromosomes. The unit differences serve as markers for chromosome loci, which are presumed to harbour specific genetic determinants, the genes. But mapping the site of the difference does not, in itself, tell us anything about the nature or dimensions of the gene. The marker is just the difference or mutation, which could be just a local scar in a much more extensive and otherwise unvarying functional unit.

The relationship between the mutation and the gene has been clarified in two different ways – first by more sophisticated genetics and, more recently, by gene cloning and sequencing. This chapter deals with the refined genetical analysis that set the scene for the application of the DNA technology, to be described in Chapter 3.

The collection of mutants

The first step in the genetic analysis of genes is the extension of the range of available mutants.

In long-standing objects of genetic study, such as *Drosophila melanogaster*, many spontaneously arising mutants have been accumulated over the

years. But if one wants many mutants of a particular kind, or to 'saturate' a particular locus with mutations, it is necessary to resort to some mutagenic treatment. X-rays and other kinds of ionizing radiation induce mutations in all organisms. Ultraviolet light, though it cannot penetrate tissues, is effective on exposed cells and is the most convenient mutagen for micro-organisms. Ionizing radiation also induces chromosome breaks and segmental rearrangements. Chemicals that react with DNA are often very potent mutagens without inducing so many chromosome breaks. The chemical mutagens in most common use are alkylating agents such as ethyl methanesulphonate (EMS) which has been used extensively on *Drosophila* as well as on micro-organisms. In general, mutagenic treatment, whether by radiation or by chemicals, needs to be sufficiently stringent to kill a substantial fraction of the treated cells if there is to be a useful proportion of mutants among the survivors.

The labour of mutant hunts can be greatly reduced if there is some method for efficient screening for mutants of a particular kind. Especially in micro-organisms, it is often possible to devise conditions in which the desired kind of mutant will survive and the non-mutated organism will not. For example, in *Escherichia coli*, penicillin can be used to kill growing cells, leaving only those unable to grow on the medium provided. In this way, large numbers of *auxotrophic* mutants, that require specific supplements in their growth medium, have been isolated.

In bacteria and yeasts the technique of *replica plating*, which is described in Fig. 2.1, has been extremely valuable for the efficient detection

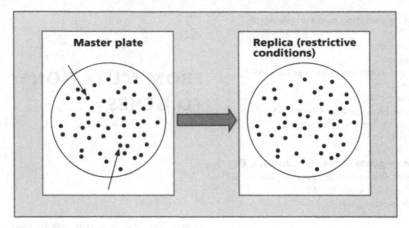

Fig. 2.1 Replica plating for the identification and isolation of conditional mutants in bacteria or yeasts. Mutagenized cells are grown into colonies on a master plate that provides permissive conditions, e.g., supplementation with ready-made metabolites such as amino acids, vitamins or nucleic acid bases, and moderate temperature. Samples of cells from the array of colonies are picked up on a replicator disc and transferred to the replica plate which has only the minimal nutrients sufficient for the wild type, or is incubated at elevated temperature. Colonies on the master plate (arrowed) that fail to replicate are putative mutants – nutritionally exacting (auxotrophs) or temperature-sensitive as the case may be.

of auxotrophic or temperature-sensitive mutant colonies that have grown up as part of an array on a master plate, but fail to grow when transferred to a replica plate that has fewer nutrients or is incubated at a higher temperature.

Mutant hunts in *Drosophila*, on the other hand, generally depend on laborious examination of individual flies or embryos. It is, however, possible to focus attention on mutations in precisely defined chromosome segments. The principle is to place chromosomes from mutagenized flies in heterozygous combination with a chromosome with a deletion covering the segment of interest. Any recessive visible or lethal mutation within that segment will thereby be revealed. One scheme is described in Fig. 2.2a. The method depends on the use of *balancer* chromosomes, of which a broad selection is available in *Drosophila*. Their main features are: (i) the presence of a recessive lethal mutation to prevent the balancer from becoming homozygous; (ii) a dominant visible marker (which may be the same as the recessive lethal) to enable flies carrying the balancer to be identified; and (iii) a long inversion or complex of inversions to prevent recombination between the balancer lethal and any mutations on the homologous chromosome with which the balancer is combined (see Fig. 1.17a). Most balancers also carry one or more recessive viable markers. The point of balancer chromosomes is that they make it easy to maintain recessive lethal mutations. For example, any chromosome 3 carrying a recessive lethal can be kept in a permanently heterozygous stock in combination with a balancer chromosome 3; such a stock will produce exclusively heterozygous progeny, since both kinds of homozygote will be inviable. The inversion in the balancer prevents the generation of homozygous-viable chromosomes by crossing over.

Mass isolation of mutants in larger organisms, such as mammals or most flowering plants, is usually too demanding of resources to be feasible.

Sorting of mutants into complementation groups

Mutant alleles that are recessive to wild type (as most mutants are) generally have either zero or reduced gene function. They are recessive because just one functional allele out of two is usually sufficient for more-or-less normal development of the diploid organism.

One can determine whether two different recessive mutants affect the same or different functions by the complementation test (Fincham, 1966). This consists of putting the two mutant genomes together in a diploid (or heterocaryon, see p. 51)

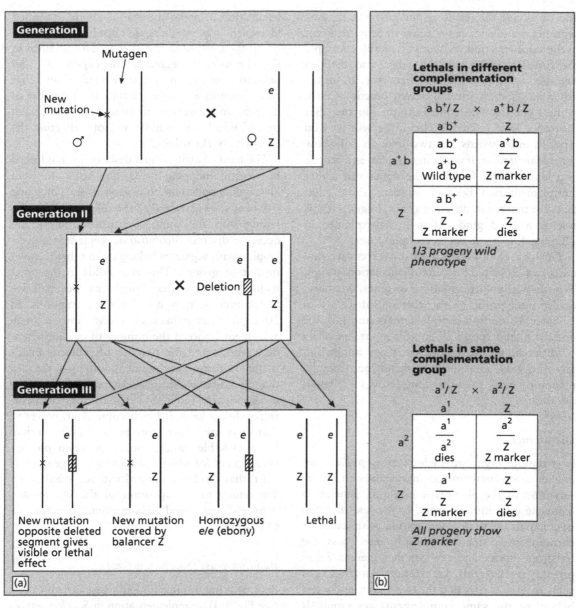

Fig. 2.2 (a) A general procedure for 'saturating' a particular segment of a *Drosophila* autosome, for example chromosome 3, with mutations. Generation I: males, treated with a mutagen (e.g., X-rays or ethylmethane sulphonate; EMS) are mated to females carrying a balancer chromosome 3. The balancer contains a recessive lethal mutation to prevent it from becoming homozygous, a dominant visible marker, here called Z (it may be the same as the recessive lethal), a recessive visible marker, for example *e* (*ebony* body colour), and an inversion to prevent effective crossing over (see Fig. 1.17a). Generation II: F_1 flies showing the Z phenotype are mated to flies in which the balancer is balanced against a chromosome 3 with the segment of interest deleted (deletion shown as box) and also marked with *e*. Generation III: If there has been a recessive mutation (x) within the segment of interest, progeny not carrying the balancer will either be missing, if x is lethal, or phenotypically distinct, if x is viable. In either case, x will be present heterozygous in all the Z non-ebony progeny.

(b) Testing of recessive lethals, obtained as shown in (a), for complementation. Mutant stocks, with the mutation in each case covered by the balancer Z, are intercrossed in many combinations. Only when the mutants complement one another (i.e., are in different genes) will any progeny not carrying the balancer be viable.

and observing the resulting phenotype. If the two mutants are allelic, in the sense of being deficient in the same function, neither will be able to supply the function lacking in the other, and the two together will give a mutant phenotype. If, on the other hand, they affect different functions, each mutant genome will contribute the function that the other lacks, and the phenotype will be wild type. In other words, the two mutants will show *complementation*. If we define the gene as a genetic unit with a single function, the simple rule is that complementation between mutants shows that they are mutant in different genes. In general this rule is a good guide, though, as explained in Chapter 4, it is subject to complications.

Through complementation tests, recessive mutants can be sorted into complementation groups, hypothetically corresponding to genes. Members of the same group (with exceptions discussed in Chapter 4) are non-complementary and give the mutant phenotype in all combinations; members of different groups complement one another to give the wild, i.e., normal, phenotype. Complementation tests take different forms in different organisms.

Drosophila

In a diploid organism such as *Drosophila*, complementation between mutants is assessed by observation of the phenotype of the F_1 generation from the inter-mutant cross. In the case of recessive lethals kept in combination with balancer chromosomes, one can simply inter-cross the balanced stocks and see whether there are any progeny not carrying the balancer. An absence of such progeny shows that the two lethals under test belong to the same complementation group. If they can make good each other's deficiencies, phenotypically wild-type flies free of the balancer chromosome should make up about one third of the progeny (Fig. 2.2b).

To take one example of an analysis of this kind, a set of 268 recessive lethal mutations was obtained in a segment of chromosome 3 consisting of about 26 polytene chromosome bands (Gausz *et al.*, 1981). The general method is explained in Fig. 2.2a. Complementation tests carried out as outlined in Fig. 2.2b showed that these 268 mutations fell into just 25 complementation groups. It seemed likely, from the numbers, that hits had been scored on all the units of essential function contained in this chromosome segment. This experiment, and numerous others of the same kind, shows that each chromosome segment contains a number of discrete and independent functional units, each one of which can mutate without affecting the functions of the others.

The kind of analysis just described is reinforced by genetic mapping on a fine scale. In both *Drosophila* and in micro-organisms such mapping (see below) has established the rule that mutations falling into the same complementation group occupy a discrete chromosome segment, not overlapping with segments belonging to other complementation groups. This rule holds in the great majority of instances though, as we shall see in Chapter 4, it is not without exceptions. In *Drosophila* numerous studies have shown a rough agreement between the number of complementation groups and the number of polytene bands within a chromosome segment, leading to the idea that the visible bands (or, perhaps more likely, the less condensed inter-band regions) may each correspond to a gene. It has become clear, however, that there are rather more units of function than distinguishable bands, especially when one includes genes (of which a surprisingly large number exist) that can be eliminated without lethal effect. We return to the question of the relationship between genes and chromosome structure in Chapter 5.

Budding yeast (Saccharomyces)

Because it has a stable diploid phase in its life cycle (see Fig. 1.5), complementation in *Saccharomyces* can, as in *Drosophila*, be observed in the progeny of sexual crosses. Each mutant to be tested has to be obtained in both mating types. Then a matrix of rows and columns can be set up on a nutrient plate, with haploid cells of all the mutants of α mating type inoculated along the rows and the equivalent series of a mating type down the columns, giving n^2 pairwise mixtures of n different mutants. Cells of different mating type fuse almost at once to give diploids which, depending on whether the mutants complement each other or

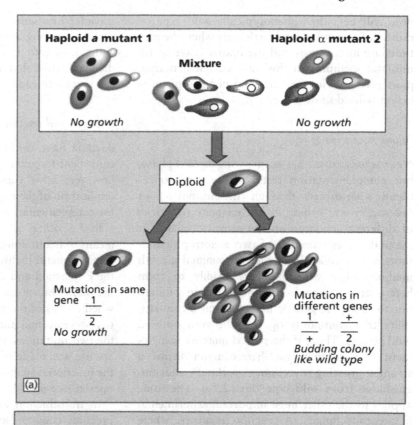

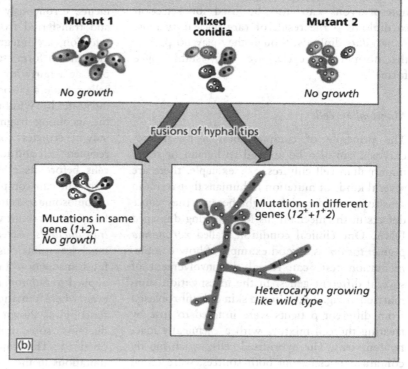

Fig. 2.3 Complementation tests on auxotrophic (nutritionally exacting) mutants of fungi. Mutant cells unable to grow by themselves without nutritional supplement are mixed on unsupplemented medium. (a) In budding yeast (*Saccharomyces cerevisiae*) haploids of opposite mating type fuse to form diploid cells. (b) In the filamentous fungus *Neurospora crassa*, asexual spores (conidia) are germinated together and hyphal fusions lead to the formation of a heterocaryon. In both cases the combination of genomes leads to ability to grow like wild type on the unsupplemented medium if the two mutants complement one another. Different mutant genomes are symbolized by white versus black cell nuclei or, in the case of the hybrid yeast diploids, as white/black sectored nuclei.

not, will be either phenotypically wild type or mutant. Scoring is particularly easy when the mutants are auxotrophs and the matrix is set up on minimal medium; in this case complementating pairs give growing colonies and non-complementation pairs do not (Fig. 2.3a).

Filamentous fungi

Neurospora crassa has no growing diploid phase, but complementation tests can be very conveniently performed through the formation of *heterocaryons* – that is, associations of nuclei of different genotypes in the same cytoplasm. If asexual spores (conidia) of two auxotrophic mutants are inoculated together on minimal growth medium, they will fuse very readily to form heterocaryons provided they are sufficiently inbred not to show the genetic heterocaryon incompatibility that commonly operates between outbred wild strains. Then, if the paired mutants complement one another, the heterocaryon forms a strongly growing mycelium superficially indistinguishable from wild type (Fig. 2.3b). The same applies to the other most important experimental filamentous fungus, *Aspergillus nidulans*, where heterocaryons can also be used for selection of diploids – the result of rare nuclear fusions. *Aspergillus* diploids, though they play no part in the normal life cycle, are fairly stable once formed.

Mammalian cells

The principle of complementation by heterocaryosis can also be applied to human or other mammalian cell cultures. For example, there are several kinds of mutation in humans that result in sensitivity to ultraviolet light because they cause defects in the mechanisms for repairing damaged DNA. One clinical condition, called *xeroderma pigmentosum*, is a good example of how complementation tests can reveal the involvement of several different genes. In the investigation summarized in Fig. 2.4, cultured skin cells (fibroblasts) from different patients were induced to fuse by treating the cell mixture with a chemically inactivated virus. The hybridized cells, which mostly contained nuclei from both sources, were tested for repair-synthesis of their DNA following ul-

traviolet irradiation. The complementation relationships among a set of 12 mutant cell lines (Fig. 2.4b) showed four complementation groups, with the implication that any one of at least four genes can mutate to cause the deficiency in DNA repair.

Bacteria and bacteriophages

Bacteria have neither full diploidy nor anything equivalent to eukaryotic heterocaryosis. They do, however, have various possibilities for partial duplication of their genomes that can be exploited for complementation testing.

In *E. coli*, F′ plasmids (see p. 40) provide a means of maintaining a second copy of a chromosome segment. If chromosomal segment carried in the F′ plasmid and the corresponding segment in the main chromosome carry different mutations, whether or not the two complement one another to restore normal function will depend on whether the two mutations are in different genes. Important use was made of F′ plasmids, for example, in the functional analysis of mutants in the *E. coli lac* operon (see p. 101).

In transduction experiments, described in the previous chapter (see p. 36), chromosome fragments, erroneously packed into phage particles and transferred from one strain to another by infection, are integrated into the recipient chromosome to form stable replicating transductants only at a fairly low frequency. More frequently, cells receive a chromosome fragment but do not integrate it. Without chromosomal integration, the incoming fragment cannot replicate, but it may nevertheless contribute gene functions to the recipient cell and to a few immediate descendant cells before its effect is diluted out. If such a fragment can complement a mutant in the main chromosome so as to restore some essential growth function, the result will be tiny colonies – *abortive transductants* – clearly distinguishable from the much less numerous large colonies resulting from full transduction (Fig. 2.5). This kind of analysis, applied to *Salmonella typhimurium*, showed that even where mutants were defective in the same function, as shown by absence of abortive transductants, some full transductants were usually obtained. The important implication was that mutations in the same gene, defined as a unit of function, were often at different sites, able to

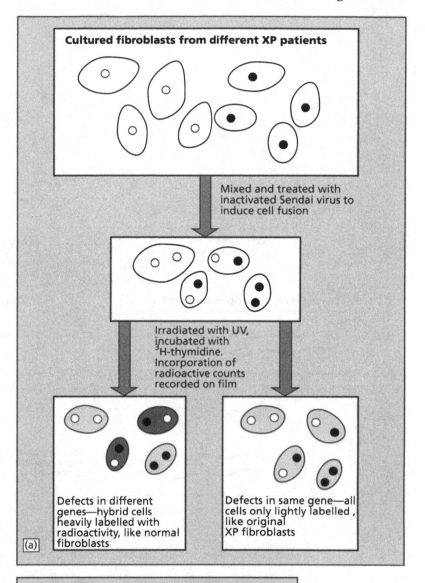

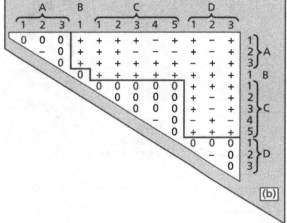

Fig. 2.4 (a) Complementation tests on cultured skin cells (fibroblasts) from different patients suffering from xeroderma pigmentosum (XP), a deficiency in ability to repair DNA following radiation damage. Cell fusion was induced by treatment with inactivated Sendai virus. The fusion products were treated with ultraviolet (UV) light and then cultured in the presence of thymidine radioactively labelled with tritium (^{3}H). Cells able to carry out normal repair of their DNA incorporate the radioactive DNA precursor, and their radioactivity is detected as silver grains in radiosensitive film superimposed on the cell preparation – shown as heavy stippling in the diagram. If the XP mutants are in different complementation groups, the hybrid cells incorporate radioactivity whereas the unhybridized cells do not.

(b) Summarized results of tests as described in (a) on cells from 12 different patients. + indicates DNA repair; 0 no detectable repair; − means that the test was not done. The 12 mutant cell lines clearly fall into 4 complementation groups A, B, C, D, with 3, 1, 5 and 3 occurrences, respectively. Data from Kraemer *et al.* (1975).

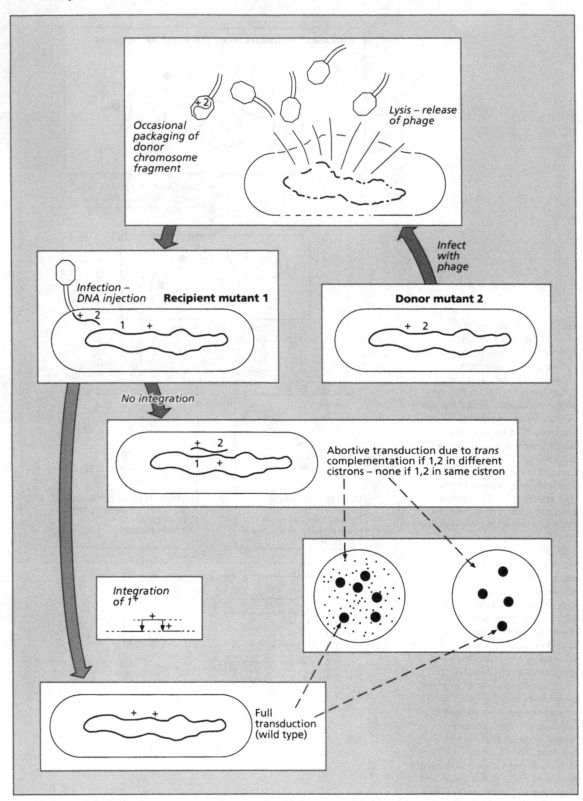

recombine to reconstitute a functional gene during the integration process.

There is a close parallel here with the type of analysis possible with T4 bacteriophage (see Figs 2.6 and 2.7), where complementation and recombination of mutants can both be assessed by mixed infections of host bacteria. We will defer discussion of recombination within genes to the next section.

An interim definition of the gene

By definition, mutants in the same complementation group cannot make good each others' deficiencies and so must have damage in common. They can be called functionally *allelic*, all with defects in the same gene. But they are no means all the same. Some show a degree of allelic complementation (see p. 96) and others not. Some have total elimination of a function, while others retain it at a reduced level, as with 'leaky' auxotrophs that grow slowly on unsupplemented medium. Some exhibit the mutant phenotype under all conditions, and others only within a certain (most often higher) range of temperature. The mutant phenotypes of some but not others may be suppressed by mutations in other genes (see p. 64). Nevertheless, they nearly always have some features in common, deviating from the wild type in similar ways though often to different extents. They can be regarded as due to different kinds of mutational damage in the same functional unit. They also map at what, at first sight at least, seems to be the same chromosome locus.

The twofold definition of the gene, as a unit that (i) resides at a unique chromosome locus and (ii) is responsible for a specific function, dates back to the early days of the chromosome theory and, with various complications to be discussed in

Chapter 4, is still useful. Another part of the classical view of the gene, that it was an indivisible unit of hereditary transmission, was undermined as the mapping of sites of mutation was pushed to a higher level of revolution.

Mapping within the gene

The detection of recombination within genes

Until the 1950s, the gene was viewed as the ultimate and indivisible unit of genetic transmission. In other words, it was thought that recombination could occur only *between* genes and never within them. If this were true, a heterozygote formed in a eukaryotic organism by crossing two allelic mutants could never (except by rare reverse mutation), form anything but mutant meiotic products. If, on the other hand, mutations fell at different sites within a gene, and recombination could occur between them, meiosis in the heterozygote will be expected to produce wild-type recombinants, albeit at a low frequency. We now know that the latter alternative is the rule.

Meiotic crossing over between the sites of allelic mutations was first demonstrated in *D. melanogaster* and, a little later, in the fungus *Aspergillus nidulans*. But the experiments that did most to establish the subdivisibility of the gene were on bacteriophage T4.

Seymour Benzer (1959) isolated a large number of T4 mutants of a class called *rII*, which were hypervirulent on *E. coli* strain B but failed to grow on *E. coli* strain K. They were sorted into two complementation groups, A and B by mixed infections; all A + B combinations lysed K cells but A + A or B + B combinations did not. But A + A or B + B mixed infections of strain B usually yielded wild-type phage particles, able to lyse strain K

Fig. 2.5 (*opposite*) The use of abortive transduction as a test for complementation and recombination between auxotrophic mutants of *Salmonella typhimurium*. Mutant 2 is infected with bacteriophage P22, and the phage released by lysis of mutant 2 cells is used to infect mutant 1. Some of the phage particles package fragments of the bacterial genome instead of phage DNA. Fragments of the mutant 2 chromosome, acquired by mutant 1 cells via phage infection, can either be integrated (at least in part) into the chromosome by homologous recombination (*full transduction*), or persist as non-replicating fragments (*abortive transduction*). If a mutant 2 fragment can supply the function lacking in mutant 1 (as can happen if 1 and 2 are in different cistrons) some recipient cells will be able to form tiny colonies on minimal medium as a result of abortive transduction. Full transduction (large colonies) can occur at some low frequency whenever mutations 1 and 2 are at different recombinable sites, even if in the same cistron. Based on data from Hartman *et al.* (1960) on histidine auxotrophs.

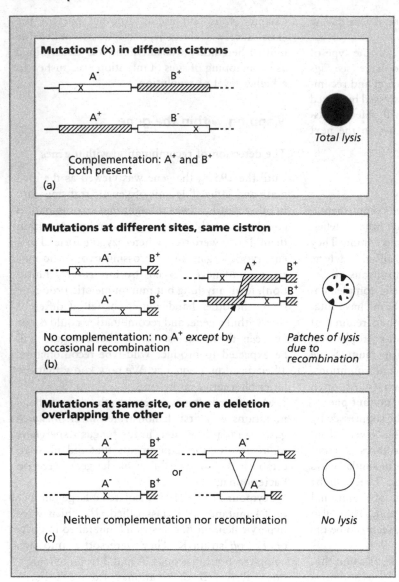

Mutations (x) in different cistrons

A⁻   B⁺

A⁺   B⁻

Complementation: A⁺ and B⁺
both present

(a)

Total lysis

Mutations at different sites, same cistron

A⁻   B⁺   A⁺   B⁺
                                    B⁺
A⁻   B⁺

No complementation: no A⁺ except by   *Patches of lysis*
occasional recombination   *due to*
                                          recombination
(b)

Mutations at same site, or one a deletion overlapping the other

A⁻   B⁺   A⁻
          or
A⁻   B⁺   A⁻

Neither complementation nor recombination   *No lysis*

(c)

Fig. 2.6 Different results of mixed infections of lawns of *Escherichia coli* strain K by pairs of bacteriophage T4 *rII* mutants, and their interpretation. Individual *rII* mutants differ from wild type in being unable to grow on *E. coli* strain K. Wild-type and mutant genes are shown as hatched and open bars, respectively, with sites of mutation as X.

(a) Complete clearing (lysis) of the mixedly infected patch, indicating that the mutants show complementation – one has lost function A and retains function B and the other has lost B but retains A.

(b) No overall lysis, indicating that the mutants have lost the same function (shown here as A), but some locally cleared spots due to the formation of occasional wild-type phage by recombination between the separable sites of mutation. The sensitivity of this test for recombination is increased if a minority of strain B cells, in which *rII* mutants can grow, is included in the lawn.

(c) No lysis at all, indicating that the sites of mutation are the same, or one is a deletion overlapping the other.

cells, with a frequency of between 0.01 and 1% – far higher than the spontaneous mutation rate to wild type of single mutants growing by themselves. Benzer concluded that different mutations were usually at different sites, separable by recombination. He was able to position a large number of sites in a linear map by reference to a set of mutants with overlapping deletions (see Figs 2.6 and 2.7, and *deletion mapping*, p. 58). A and B mutants mapped in different though contiguous segments.

On the basis of these results, Benzer proposed a new term, *cistron*, to describe the functional gene. It was defined by the comparison of two different ways of combining two linked mutational differences, call them *a*/+ and *b*/+. The *cis* arrangement is *a b*/+ + and the *trans* arrangement is *a* +/+ *b*. If *a* and *b* fall within independently functioning genes (cistrons), *cis* and *trans* will be phenotypically identical – more or less wild type, depending on whether the mutants are completely recessive. If they fall in the same cistron, *cis* will

Fig. 2.7 (a) The principles of Seymour Benzer's analysis on the relationships between T4 *rII* mutants. The symbols in the matrix indicate the three different results of mixed infections of *Escherichia coli* K cells by pairs of *rII* mutants: (i) ● full complementation, as in Fig. 2.6a, giving a cleared patch on the K cell lawn; (ii) + no complementation, but formation of some wild-type phage particles by recombination, as in Fig. 2.6b; (iii) − neither complementation nor recombination, as in Fig. 2.6c.

(b) A map based on the results shown in (a). Mutants 1–10 and 15–21 are apparently point mutants; 5 is at the same site as 6, and 18 and 19 at the same site as 20. Mutants 11, 12, 13, 14, 22 and 23 are deletions overlapping sets of point mutations as shown, and they partially define the sequence of the mutations; the positions within the groups (1, 2, 3), (8, 9) and (18, 21) are not defined. Mutants 1–13 form one complementation group (cistron) and mutants 15–23 another. Mutant 14 is a deletion overlapping both groups and complementing neither. The overall picture is one of two adjacent cistrons (genes), with many individually mutable and recombinable sites within each. The numbering of the mutants is arbitrary in this selective presentation. Based on Benzer (1959).

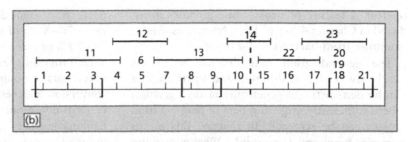

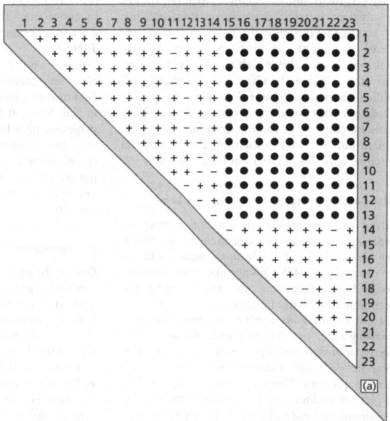

approximate to wild type and *trans* will give the mutant phenotype. In practice, the examination of just the *trans* phenotype – the simple complementation test – usually suffices. Non-complementing mutants are assigned to the same cistron.

The cistron concept was soon found to apply to all situations in which fine-structure mapping was possible. In the analysis of auxotrophic mutants of *E. coli* and *S. typhimurium* transduction was used to test for both complementation and recombina-

tion. A notable example was the analysis of histidine-requiring auxotrophs of *S. typhimurium*, referred to in Fig. 2.5. Here a large number of mutually recombinable mutations were shown to fall into nine different complementation groups or cistrons, each deficient in one of the enzymes catalysing the different steps in histidine biosynthesis. We should note in parentheses that these cistrons, occupying adjacent, non-overlapping segments in the genetic map, are part of the same *operon*, a feature of prokaryotic genomes introduced in Chapter 4 (see p. 101). Numerous similar examples from bacteria could be cited.

The generalization that different mutations nearly always fall at different and separable sites can be demonstrated genetically in any organism where sufficiently large numbers of progeny from crosses can be screened. In practice, this means bacteria, fungi and *Drosophila*. What constitutes a sufficient number depends on the frequency of intra-gene recombination, which varies very greatly from one organism to another. In *Saccharomyces* it can be as high as several per cent, so analysis of a few hundred meiotic tetrads will suffice. In multicellular eukaryotes, one usually needs progeny numbering tens of thousands to have much hope of picking up a meiotic recombinant between sites in the same gene – numbers obtainable in *Drosophila* but hardly in mice.

Mapping by reference to flanking markers

Given that mutational sites within a gene are separable by recombination, the next question is whether such sites fall in a linear sequence that is an integral part of the linkage map of the chromosome. If this were the case, and assuming that recombination within genes occurs in the same way as recombination between genes, that is by crossing over, one would expect that intra-genic recombinants would show reciprocal recombination between any markers closely flanking the gene in question. This does tend to be true and, as explained in Fig. 2.8, it provides a method for obtaining the order of sites within a gene relative to the flanking markers.

However, this kind of analysis is often somewhat complicated by the unexpectedly large proportion of intra-genic recombinants that retain the parental combinations of flanking markers. According to simple crossover theory, such products would be expected only as a result of double crossing over which, with closely placed markers, should be exceedingly rare. The answer to this paradox is that recombination within genes is generally due, not to crossing over as such, but rather to *conversion* events – non-reciprocal transfers of patches of gene sequence from one chromatid to its homologue. Conversion tends to

occur in the immediate vicinity of crossovers, but up to 70–80% of conversions in *Drosophila* and 50–70% of those in fungi occur without crossing over. Provided that conversion-associated crossovers usually occur immediately adjacent to the conversion patches (as appears to be the case in *Drosophila* but not so consistently in fungi), flanking markers can still be used to define the order of mutant sites undergoing recombination by conversion. The principle is explained in Fig. 2.8 and an example from *Drosophila* is shown in Table 2.1.

Gene conversion and its relation to mechanisms of recombination is a fascinating but very complex field of study, and it will not be discussed further in this book. It becomes significant for genetic mapping only when the sites being recombined are close to each other and so necessarily close to the recombinational event itself, which turns out to be not quite the clean break–rejoin process that was inferred from classical mapping with well-spaced markers.

Recombination frequency

One might expect to be able to map mutational sites within genes on the basis of distances between them deduced from recombination frequencies. Even if recombinants are due to conversion, the chance of recombination must depend on the conversion patch including one site and not the other, and this should become the more probable the wider the separation between the sites. But although recombination frequency within genes is a rough guide to distance it is far from being a reliable one, probably because frequencies of conversion are affected by the nature of the mutational sites that are being converted. The most decisive way of mapping sites within genes is through the use of deletions.

Deletion mapping

Most induced mutations can be classified into two groups: point mutations and deletions. The former, known from molecular analysis to be mostly single base-pair substitutions in the DNA, are generally capable of mutating back to wild type. The other characteristic feature of point

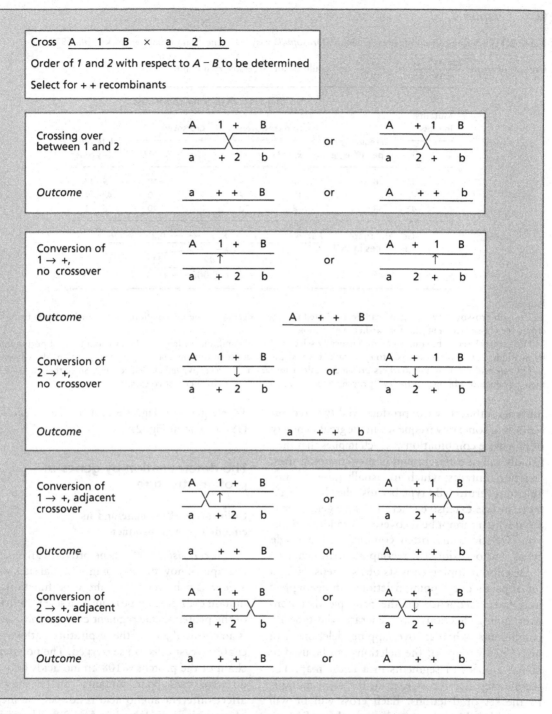

Fig. 2.8 The establishment of the order of mutational sites within a gene by the use of flanking markers, and the complication introduced by crossover-associated conversion. The mutant sites within the gene are shown as *1* and *2*, with their corresponding wild-type sites both shown as +. Flanking markers are shown as *A*, *a* on the left and *B*, *b* on the right, with one parental chromosome <u>*A 1 B*</u> and the other <u>*a 2 b*</u>. Wild-type (++) recombinants within the gene will, if all due to single crossovers, all be either <u>*A b*</u> or <u>*a B*</u>, depending on whether site *1* is to the right or to the left of site *2*. If ++ recombinants arise by conversion events, either *1* → + or *2* → +, the method still works provided that the converted gene segments are on the same side of the gene as the associated crossovers. This assumption appears to hold in *Drosophila* (see Table 2.1) but not always in fungi.

Table 2.1 Ordering of mutant sites within the *Drosophila rosy* (*ry*) gene. Data from Chovnick *et al.*, 1971

Crosses of form: $\dfrac{kar\ ry^x\ l26}{+\ ry^y\ +} \times \dfrac{+\ ry^*\ +}{+\ ry^*\ +}$ (*ry** could be any recessive *ry* mutant)

| Mutants in cross | | Frequency ry^+ per 10^6 eggs | No. of ry^+ in each flanking marker class | | | | Inferred x–y order |
| | | | Non-crossovers | | Crossovers | | |
x	y		kar l26	+ +	kar +	+ l26	
5	42	26	7	3	0	10	5–42
41	42	23	18	7	5	0	42–41
5	41	58	8	30	0	40	5–41

Genetic map: (distances in cM)

$$kar \longleftarrow 0.3 \longrightarrow ry \longleftarrow 0.2 \longrightarrow l26$$

$$\boxed{5 \qquad 42 \qquad 41}$$

$$\longleftarrow c.\ 0.006 \longrightarrow$$

Notes

1 The non-crossover ry^+ progeny can be attributed to gene conversion. The information on x–y order comes from the crossover classes (see Fig. 2.8 for further explanation).

2 The rare ry^+ recombinants were automatically selected from the millions of laid eggs by incorporation of purine into the fly medium. Purine is highly toxic to *ry* mutants because they lack the purine-metabolizing enzyme xanthine dehydrogenase.

3 The *kar* and *l26* flanking markers are an eye colour mutant and a recessive lethal, respectively. Since they are recessive, their presence or absence in the ry^+ progeny had to be determined by further test-crosses.

mutants is that they can produce wild-type recombinants at some low frequency in the great majority of pair-wise combinations, which implies that they usually fall at different and non-overlapping sites. Deletion mutants, which are usually rarer, are unable to revert to wild type because they have lost a more or less extensive tract of unique genetic material which cannot be recovered by random mutation. Deletions are often confined within single genes, but occasionally overlap adjacent genes.

Deletion mapping consists of two steps. First, a set of partly overlapping deletions can be mapped in a linear sequence on the principle that non-overlapping deletions can generate wild-type recombinants whereas overlapping deletions can not. The overlaps of the deletions can be used to define a series of segments in a linear map. The second step is to cross each of the point mutants to the set of deletions. Each cross will or will not yield wild-type recombinants depending on whether the site of the point mutation falls outside or inside the segment deleted in the other parent. On this basis, each point mutation can be placed within one of the segments defined by the deletion overlaps.

The principle is illustrated for the bacteriophage

T4 *rII* genes in Fig. 2.6 and for the *S. cerevisiae* *CYC1* gene in Fig. 2.9.

The determination by genes of protein structure

Colinearity of the gene and its encoded protein product

The *S. cerevisiae CYC1* gene provides an excellent example of how the map of mutational sites within a gene can be related to the gene function. The mutant *cyc1* phenotype is due to a severe deficiency of the protein-heme pigment cytochrome c, which is an essential part of the respiratory pathway that enables yeast cells to use oxygen. The polypeptide chain of the protein is 108 amino acids long, and different *cyc1* point mutants can be shown to affect different amino acid residues. The method of analysis is explained in Fig. 2.9. The striking result is that the sequence of mutant sites in the gene map is the same as the sequence in the polypeptide chain of the amino acid residues affected by the mutations.

A similar conclusion has been reached for numerous other gene–protein relationships. An ex-

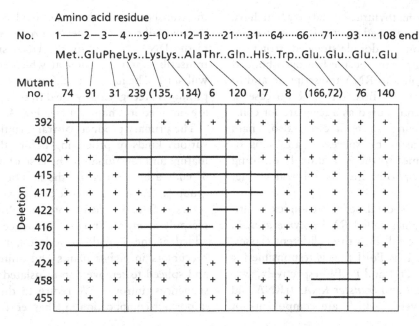

Fig. 2.9 Establishment of the sequence of sites of mutation in a series of *Saccharomyces cerevisiae cyc1* point mutants by crossing to a set of deletion mutants. Point mutants are listed in the matrix horizontally and deletion mutants vertically; + indicates formation of wild-type recombinants. The horizontal lines show the extents of the deletions. The top line shows the positions within the cytochrome c sequence of the amino acid residues affected by the point mutations. Most of the primary mutants were deficient in cytochrome c because of mutation to chain termination, and the codon affected in each point mutant was determined after a second mutation had restored a different amino acid in that position. The over-representation of glutamate and lysine mutants in the collection is due to the fact that the codons for these amino acids (glutamate GAG, GAA; lysine AAG, AAA) can mutate to chain termination (UAG, UAA) by single base changes (see Table 2.2). From Sherman *et al.* (1975).

ample from *Drosophila* is the *rosy* gene (see Table 2.1), in which mutations affect eye colour by reducing or eliminating the activity of the enzyme xanthine dehydrogenase (XDH). Here again, the positioning of the mutational sites within the gene map mirrors the sequence of the affected amino acid residues in the XDH polypeptide chain, so far as these have been determined.

This colinear relationship between specific genes and specific polypeptides suggested very strongly that genes (or at least some genes) were linear codes for the sequences of amino acid residues in specific proteins – one gene for each kind of polypeptide.

Biochemistry of polypeptide synthesis

By the time the coding relationship between genes and the polypeptide chains of proteins had been substantiated by the kind of evidence just re-

viewed, the biochemical mechanism of protein synthesis from amino acids had been largely worked out. This book is not concerned with biochemical mechanism as such, but at this point it may be as well to summarize, albeit without evidence, the basic steps whereby information encoded in DNA is translated into amino acid sequence.

Transcription

The sequences of amino acids in polypeptides are determined, not by DNA directly, but by *messenger ribonucleic acid* (mRNA) molecules that are transcribed from DNA. RNA resembles DNA in consisting of chains of nucleotide residues but differs in two respects. Firstly, the sugar component is ribose rather than deoxyribose; secondly, the pyrimidine base thymine is replaced by uracil. The two bases are closely related chemically –

thymine is just 5-methyluracil – and they both have adenine as their preferred partner for base-pairing in the formation of double-stranded structure. The same principle of specific hydrogen-bonded base-pairing applies to RNA transcription and to DNA replication (see Box 1.1); in both cases a single DNA strand is used as a template for complementary copying. In both cases, also, chain elongation proceeds by successive additions of nucleotide residues to the 3'-hydroxyl end using nucleoside 5'-triphosphates as substrates (Fig. 2.10).

Transcription is catalysed by complex multi-component enzymes called *RNA polymerases*. In eukaryotes, these fall into three different classes, PolI, II and III. It is PolII that is responsible for making mRNA. PolI and PolIII, respectively, synthesize *ribosomal* and *transfer RNAs* (rRNA and tRNA, respectively), which are components of the translation machinery outlined below. These enzymes all recognize specific *promoter* sequences in the DNA that specify the DNA strand to be transcribed and the point at which transcription will start. The important differences between the promoters recognized by the different polymerases are outlined in Chapter 4 (see Fig. 4.8).

The primary products of transcription undergo various kinds of processing before they become mature and functional. In the case of mRNAs, the 5' ends are stabilized by the addition of a 'cap' group (a guanine nucleoside derivative attached in reversed 5'-to-5' orientation), and 'tails' of simple poly-adenylate (poly-A) sequence are usually added at the 3' end. In addition, most mRNA precursors in higher plants and animals are cut and spliced to remove non-translated intervening sequences (introns). We return to this extremely important complication in Chapter 4.

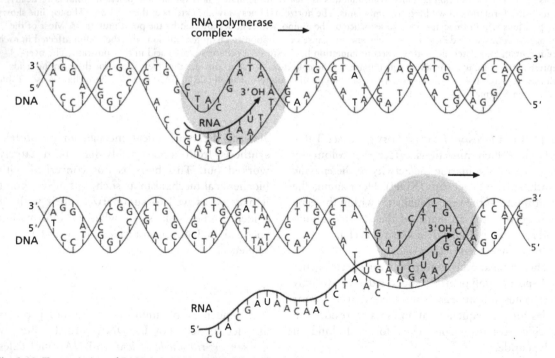

Fig. 2.10 Transcription of DNA into RNA. A simplified sketch of synthesis of RNA using DNA as template. The DNA duplex is locally unwound and just one of its strands is transcribed into RNA. Specific promoter sequences in the DNA bind the RNA polymerase and other protein components of the initiation complex (see Chapter 7). Ribonucleotide units with the bases adenine (A), guanine (G), uracil (U) or cytosine (C), supplied in the form of nucleoside 5'-triphosphates (ATP, GTP, UTP or CTP), are successively transferred to the 3'-hydroxyl end of the RNA chain. The sequence of the RNA is determined by that of the template DNA strand by complementary base pairing: RNA bases A, G, U and C are brought in opposite DNA bases T, C, A and G, respectively.

Translation

mRNA molecules are translated into amino acid (polypeptide) sequence by ribosomes, which are two-subunit protein–rRNA complexes. The larger and smaller ribosomal subunits each contains its own special molecule of rRNA – large and small rRNA, respectively (see Fig. 4.8a) – and its own set of ribosomal proteins.

Ribosomes track along the mRNA molecule in a 5′-to-3′ direction, and read off the sequence of mRNA bases in threes, using a set of tRNA molecules as translation keys. Each sequence of three, called a *codon*, specifies one of the 20 protein amino acids, except for the three triplets, UAA, UAG and UGA, which code for no amino acid and signal polypeptide chain termination. The sequence AUG codes for methionine and is also the translation initiation signal – the ribosomes usually start at the first AUG downstream of the cap site at the 5′ terminus of the messenger.

Each kind of tRNA has two kinds of specificity. It is recognized by a particular aminoacyl-tRNA synthetase, which catalyses the attachment to it of a specific amino acid. It also has a specific exposed trinucleotide sequence, the *anticodon*, which enables it to recognize one or more codons, all encoding that same amino acid (Table 2.2). The combination of mRNA codon and ribosome binds aminoacyl-tRNA of a kind determined by the

Table 2.2 mRNA codons and tRNA anticodons

Amino acid	Codons	Anticodons	Amino acid	Codons	Anticodons	Amino acid	Codons	Anticodons
Phenylalanine	UUU UUC	GAA	Proline	CCU CCC	IGG	Aspartic acid	GAU GAC	GUC
Leucine	UUA	UAG		CCA	UGG	Glutamic acid	GAA	UUC
	UUG	CAA		CCG	CGG		GAG	CUC
	CUU CUC	IAG	Threonine	ACU ACC	IGU	Cysteine	UGU UGC	GCA
	CUA	UAG		ACA				
	CUG	CAG		ACG		Tryptophan	UGG	CCA
Isoleucine	AUU AUC	IAU	Alanine	GCU GCC	IGC	Arginine	CGU CGC	ICG
	AUA	UAU		GCA	UGC		CGA	UCG
Methionine	AUG	CAU		GCG			CGG	CCG
Valine	GUU GUC	IAC	Tyrosine	UAU UAC	GUA		AGA	UCU
							AGG	CCU
	GUA	UAC	Histidine	CAU CAC	GUG	Glycine	GGU GGC	GCC
	GUG	CAC						
Serine	UCU UCC	IGA	Glutamine	CAA	UUG		GGA	UCC
				CAG	CUG		GGG	CCC
	UCA	UGA	Asparagine	AAU AAC	GUU			
	UCG	CGA						
	AGU AGC	GCU	Lysine	AAA	UUU			
				AAG	CUU			

The anticodons are those listed by Jukes *et al.* (1987) as having been identified in eukaryotic tRNAs.

Both codons and anticodons are written 5′-to-3′, left-to-right. Since they pair in inverted orientation, one must read the anticodon sequences backwards in order to see the complementary fits to the corresponding codons.

I = inosine, the nucleoside of hypoxanthine, an adenine derivative with the same hydrogen-bonding specificity as guanine.

Note that many tRNAs can serve two different codons, usually because G or I in the first position of their anticodons will pair satisfactorily with U (two hydrogen bonds) as well as with C (three hydrogen bonds) in the third codon position.

Anticodon bases are often modified, especially by methylation. Uracil is sometimes joined in an unusual way to the ribose moiety to make pseudouridine instead of uridine. These various modifications do not affect base-pairing specificity and are not shown in this table.

complementary base-pairing of codon and anticodon. Thus the tRNAs deliver amino acids to the growing end of the polypeptide chain in a sequence corresponding to the string of codons in the messenger.

The whole system depends crucially on the maintenance of the 'reading frame' – the system of reading in threes that defines where codons begin and end. There are three ways, or phases, in which the messenger sequence can be read and only one is correct. The correct phase is set by the initiating AUG codon. Beyond the AUG, the coding sequence, or *open reading frame* as it is called, continues until it runs into one of the three termination codons. Coding sequences for proteins can be recognized in DNA sequences as open reading frames extending over hundreds or even thousands of bases; in a random sequence, termination codons would be expected to occur every 50 to 100 bases.

The deletion or addition, within an open reading frame, of any number of mRNA nucleotide residues other than a multiple of three will result in a change of phase – a *frameshift* – with consequent mistranslation and probable early termination.

A diagrammatically simplified picture of messenger translation is shown in Fig. 2.11.

Not all genes encode proteins

The fine-structure mapping of genes, together with the analysis of the effects of gene mutation on protein structure, consolidated the concept of 'one gene–one polypeptide chain'. The same biochemistry that showed how this could work also made it clear that the determination of amino acid sequence could not be the only function of genes.

The rRNAs and tRNAs, which are universally involved in protein synthesis, are also transcribed from DNA, and the genes involved have indeed been identified. It is particularly easy to detect rRNA genes by nucleic acid probing, as described in the next chapter, but, like the protein-encoding genes, they first showed themselves through effects of their mutations. Since they are present in the genome in many clustered copies, concentrated in eukaryotic genomes at the chromosome nucleolus organizer regions (see p. 132), it takes the deletion of a sizeable block of them to make an impact on the phenotype. The *bobbed* (*bb*) series of *Drosophila* mutants have deletions of this kind. They have a deficiency in protein synthesis, the most obvious symptom of which is a shortening of the bristles of the thorax and abdomen.

tRNA genes are also generally multicopy, though not to the same extent as rRNA genes. They were originally identified through mutations turning them into dominant *suppressors* of the effects of mutations in protein-encoding sequences. Changes in tRNA anticodons can result in codon misreading; one amino acid is inserted in place of another, or in response to a chain termination codon. Such misreading can often restore function to an otherwise crippling mutant codon in an essential open reading frame. Suppressor mutants are viable only because tRNA genes are multicopy; non-mutant copies remain to carry out the normal decoding function.

Conclusions

Until the 1950s the gene was regarded both as a functional unit and as indivisible by recombination. Chromosomal crossing over was thought always to occur between genes, never within them. The functional unit was defined then as now by the criterion of non-complementation of allelic mutants, and alleles were supposed to segregate from one another at meiosis without recombination.

This concept of the gene as an indivisible unit of function was undermined first by large-progeny analysis in *Drosophila* and, even more decisively, in microbial systems that permitted the screening of still larger numbers. The classical gene was replaced by the cistron, a linear array of individually mutable and recombinable sites that together constituted a single integrated unit of function.

As the effects of gene mutation on protein structure began to be analysed, it became evident that the linear array of sites in the gene-as-cistron could, and perhaps generally did, form a code for the linear arrangement of amino acid residues in a polypeptide chain. This conclusion fitted well with the evidence from transformation experiments that the gene consisted of DNA, and the growing body of biochemical evidence that amino acid sequence was determined by mRNA molecules that were in turn transcribed from DNA.

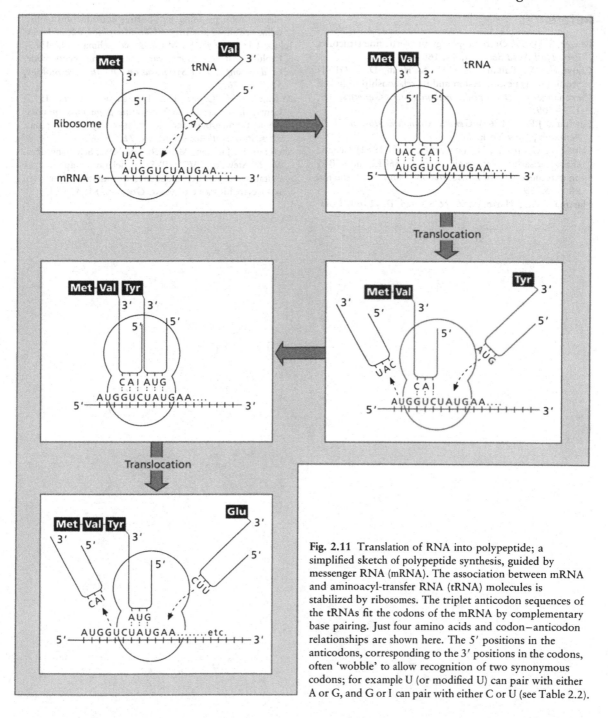

Fig. 2.11 Translation of RNA into polypeptide; a simplified sketch of polypeptide synthesis, guided by messenger RNA (mRNA). The association between mRNA and aminoacyl-transfer RNA (tRNA) molecules is stabilized by ribosomes. The triplet anticodon sequences of the tRNAs fit the codons of the mRNA by complementary base pairing. Just four amino acids and codon–anticodon relationships are shown here. The 5′ positions in the anticodons, corresponding to the 3′ positions in the codons, often 'wobble' to allow recognition of two synonymous codons; for example U (or modified U) can pair with either A or G, and G or I can pair with either C or U (see Table 2.2).

What was missing from this picture was any first-hand description of the gene itself. The gene was still something inferred from the effects of its mutation on other parts of the living system.

The next chapter describes how this situation was remedied through the development of techniques for identifying and isolating specific DNA sequences.

References

Benzer, S. (1959) On the topology of genetic fine structure. *Proc Natl Acad Sci USA*, **45**, 1607–20.

Chovnick, A., Ballantyne, G.H. & Holm, D.G. (1971) Studies on gene conversion and its relationship to linked exchange in *Drosophila melanogaster*. *Genetics*, **69**, 179–209.

Fincham, J.R.S. (1966) *Genetic Complementation*. W.A. Benjamin, New York.

Gausz, J., Gyurovics, H., Beneze, G. *et al.* (1981) Genetic characterisation of the region between 86F12 and 87B15 on chromosome 3 of *Drosophila melanogaster*. *Genetics*, **98**, 775–89.

Hartman, P.E., Hartman, Z. & Serman, D. (1960) Complementation mapping by abortive transduction of histidine requiring mutants. *J Gen Microbiol*, **22**, 354–68.

Jukes, T.H., Osawa, S., Muto, A. & Lehman, N. (1987) Evolution of anticodons: variations in the genetic code. *Cold Spring Harbor Symposia on Quantitative Biology*, **52**, 769–76.

Kraemer, K.H., Coon, H.G., Petinga, R.A. *et al.* (1975) Genetic heterogeneity in *Xeroderma pigmentosum*: complementation groups and their relationship to DNA repair rates. *Proc Natl Acad Sci USA*, **72**, 59–63.

Sherman, F., Jackson, M., Liebman, S.W., Schweingruber, M. & Stewart, J.W. (1975) A deletion map of *cyc1* mutants and its correspondence to mutationally altered iso-1 cytochromes c of yeast. *Genetics*, **81**, 51–73.

3

. . . .

THE GENE AS DNA SEQUENCE

Characterization of DNA fragments

Restriction endonucleases

For a long time research on DNA was stalled by the difficulty of showing that it was anything other than a repetitive spaghetti-like mass. Unlike any respectable chemical, it appeared to have no fixed molecular weight and there was no known way of fractionating it into distinct components.

The major breakthrough that made DNA analysable was the discovery of highly specific enzymes – restriction endonucleases – for cutting it into defined pieces. Such enzymes are produced by many different species and strains of bacteria, seemingly with the function of degrading DNA from alien organisms. Each bacterial strain tends to have its own enzyme that will cut DNA at a specific sequence, usually a palindrome of 4 or 6 base-pairs (bp). The bacterium protects its own DNA from being cut by its own enzyme by specific methylation of the potential cutting sites. This is known as a restriction-modification system; the access of alien DNA is restricted and indigenous DNA is protected by modification.

The enormous importance of restriction endonucleases depends on the relative infrequency of their cutting sites. If DNA had equal frequencies of all four bases in random sequence, one would expect an enzyme that recognizes a 6 bp sequence (a 'six-cutter') to cut DNA on average once in 4^6 = 4116 bp. In practice, this is usually not far from the average fragment size, with fragments falling predominantly within the range 500 bp to 20 000 bp (20 kb). The array of sizes is constant and characteristic of the source of the DNA.

Table 3.1 lists some of the more widely used restriction endonucleases and the restriction sites at which they cut. As shown in the table, most of them make staggered cuts, leaving short single-stranded overhangs which have the potential for mutual base-pairing ('stickiness').

Separating and sizing DNA fragments

The practical usefulness of restriction enzymes depended on the availability of a relatively simple way of separating and sizing the fragments that they generate. This is provided by the technique of *gel electrophoresis*, in which samples of restriction digests are made to migrate in an electric field through a gel, usually made from the complex polysaccharide agarose buffered at a somewhat alkaline pH. The DNA fragments, which carry a virtually uniform negative charge per unit mass, migrate towards the positive end of the gradient at a rate determined by the degree to which they are retarded by the pores of the gel. The larger the size of the fragments in relation to the pore size of

Table 3.1 The action of some restriction endonucleases

Enzyme	Target sequence*	Cut ends		
*Hind*III†	... AAGCTT ...	... A	+	AGCTT ...
	... TTCGAA ...	... TTCGA		A ...
*Eco*RI	... GAATTC ...	... G	+	AATTC ...
	... CTTAAG ...	... CTTAA		G ...
*Bam*HI	... GGATCC ...	... G	+	GATCC ...
	... CCTAGG ...	... CCTAG		G ...
*Pst*I	... CTGCAG ...	... CTGCA	+	G ...
	... GACGTC ...	... G		ACGTC ...
*Hha*I	... GCGC ...	... GCG	+	C ...
	... CGCG ...	... C		GCG ...
*Hpa*II, *Msp*I‡	... CCGG ...	... C	+	CGG ...
	... GGCC ...	... GGC		C ...
*Alu*I	... AGCT ...	... AG	+	CT ...
	... TCGA ...	... TC		GA ...
*Not*I	... GCGGCCGC ...	... GC	+	GGCCGC ...
	... CGCCGGCG ...	... CGCCGG		CG ...

* Sequences are written, left-to right, 5′ to 3′ in the upper strand and 3′ to 5′ in the lower.
† The enzymes are named after the bacteria that make them. Thus *Hind*III is the third endonuclease isolated from *Hemophilus influenzae* strain d.
‡ *Hpa*II is restricted by methylation of the inner C and *Msp*I by methylation of the outer C (see p. 156).

the gel the more slowly they will move. Increasing the agarose concentration, and thus decreasing the pore size, retards larger fragments more drastically than smaller ones. Within a certain size range, which will depend on the agarose concentration, there is an approximately logarithmic relationship between fragment size and distance moved. Mixtures of fragments of known sizes are run in parallel with the experimental samples to provide points for the construction of a standard curve for size estimation.

Restriction digestion of the DNA of a simple genome, for example of a relatively small virus, yields a limited number of different restriction fragments, usually all of different sizes. Gel electrophoresis then resolves the mixture into discrete bands that can be easily visualized and photographed after staining with the acridine drug ethidium bromide. This reagent binds into the stack of DNA base-pairs to form a complex that fluoresces intensely in ultraviolet light.

However, any restriction digest of the total genomic DNA of a cellular organism, even a bacterium, contains so many fragments that they merge on the gel into a smear, with perhaps some bands representing repetitive sequences standing out faintly above the ethidium-stained background. Getting information from such complex digests requires some means of probing for specific components.

Probing for specific sequences

Although there is little to distinguish different double-stranded DNA fragments in their general chemical properties, the two complementary strands of any DNA duplex above a certain length – 20 to 30 bp is often sufficient – can recognize each other with surprising efficiency. If 'melted' apart by heat or alkali treatment, they will, if allowed to reanneal slowly, do so with great discrimination even in the presence of a large excess of other single-stranded fragments. This is due to the powerful stabilizing effect of a continuous series of mutually compatible base pairs. It is not only a question of the cumulative effect of the individually weak hydrogen bonds; there is also a strong additional stability resulting from the stacking of adjacent hydrogen-bonded base-pairs along the axis of the duplex. The same principle applies to mutually complementary single-stranded DNA–RNA associations, such as between an RNA transcript and the DNA strand from which it was transcribed.

The great specificity of mutual binding of complementary nucleic acid strands means that one can use an appropriately (usually radioactively) labelled single DNA strand as a probe to find the complementary sequence in a complex mixture. This principle can be applied to restriction digests through the procedure known as *Southern blotting* (after the inventor E.M. Southern). The DNA is fractionated on a gel, made single stranded by soaking the gel in alkali, and then the whole array of bands is blotted with minimal disturbance on to a membrane made from a material (usually nitrocellulose or nylon) to which it will stick (Fig. 3.1). The binding of the DNA to the membrane is then made chemically secure either by brief ultraviolet irradiation (for nylon) or by baking (for nitrocellulose). The membrane is bathed for several hours in a solution containing the radioactively labelled probe to allow *hybridization* (duplex formation) between the probe molecules and the membrane-bound target molecules. Unspecifically adsorbed radioactivity is removed by washing, and the membrane is dried and exposed to X-ray film to locate any radioactively labelled DNA bands.

The temperature and salt concentration at the

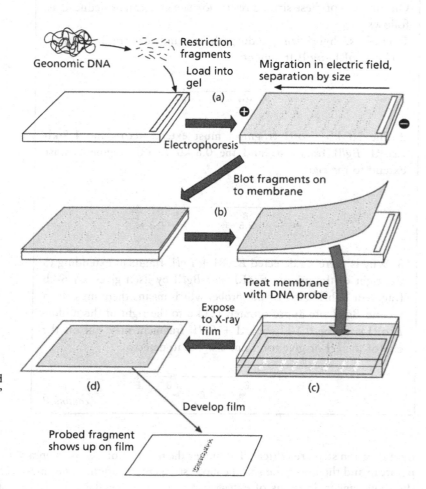

Fig. 3.1 The Southern procedure for blotting DNA fragments from electrophoretic gels and locating them on the blot by radioactive DNA probes. (a) Electrophoresis of DNA restriction digests in agarose gel. (b) Transfer to nitrocellulose or nylon membrane after denaturing to single-stranded form with alkali. (c) 'Hybridization' of membrane with radioactively labelled single-strand DNA corresponding in sequence to the fragment sought. (d) Location of hybridized sequences by autoradiography.

Box 3.1 An example of restriction-site mapping – the rabbit β-globin gene. Data from Jeffreys & Flavell (1977)

. . . .

Rabbit DNA was singly or doubly-digested by restriction endonucleases, and the fragments were separated by size on agarose gels, blotted on to membranes and hybridized to radioactive rabbit β-globin cDNA. The sizes of the fragments recognized by the probe were estimated by comparison with standards, with the following results.

Single digests	kb	Double digests	kb
*Eco*RI	2.6, 0.8	*Eco*RI + *Pst*I	1.3, 0.8
*Pst*I	6.3	*Pst*I + *Kpn*I	3.6
*Kpn*I	5.1	*Eco*RI + *Bgl*II	1.5
*Bgl*II	1.6	*Kpn*I + *Eco*RI	2.5, 0.8

On the basis of these sizes, a restriction site map can be deduced as follows.

1 The DNA hybridizing to the probe is mainly or entirely within a 1.5 kb *Eco*RI–*Bgl*II fragment:

2 The 2.6-kb *Eco*RI fragment must extend across the 1.5-kb *Eco*RI–*Bgl*II fragment, and the 0.8-kb *Eco*RI fragment must extend to the right:

3 Why is there no detected *Eco*RI + *Bgl*II fragment extending to the right of the middle *Eco*RI site? *Bgl*II by itself gives a 1.6-kb fragment hybridizing to the probe, which means there must be a second *Bgl*II site approximately 0.1 kb to the right of the middle *Eco*RI site; the 0.1-kb *Eco*RI + *Bgl*II fragment may run off the end of the gel or give too faint a signal to detect:

E ————— B ·········· E B ——— E
          ◄·········1.6·········►

continued

hybridization step are critical. The higher the temperature and the lower the salt the more stringent the requirement, in terms of extent and accuracy of complementary matching, for stable hybridization. The more complex the mixture under analysis the more stringent are the conditions

Box 3.1 *Continued*

4 The 2.5-kb *Kpn*I + *Eco*RI fragment must extend across the probed sequence to the left of the middle *Eco*RI site; the 0.8-kb *Kpn*I + *Eco*RI fragment is due to *Eco*RI alone:

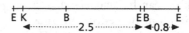

5 Since *Kpn*I alone gives a 5.1-kb probed fragment, we can place another *Kpn*I site site to the right.

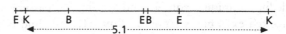

6 The *Pst*I-*Eco*RI double digest gives a 1.3-kb probed fragment which must extend to the left of the middle *Eco*RI site; again the 0.8-kb fragment is due just to *Eco*RI:

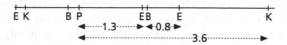

The 3.6-kb fragment given by *Pst*I + *Kpn*I gives independent confirmation of the position of the *Pst*I site.

7 *Pst*I alone gives a 6.3-kb probed fragment, which enables us to position a second *Pst*I site far to the right.

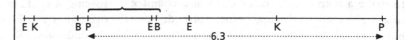

The evidence presented here shows that the probed sequence must be confined to the bracketed segment of the map. Further analysis, using other restriction enzymes, delimited it more closely (see Fig. 4.6). Note that only sequences hybridizing to the cDNA probe are detected, so no more than one site cut by a particular enzyme on each side of the probed sequence could be identified in this analysis.

required to pick out a unique sequence. The larger the genome, the longer a sequence (and the probe for it) has to be in order to be uniquely identified.

The probe length and the hybridization conditions are chosen to suit the objective in view.

The key step in this kind of analysis is obviously

obtaining the appropriate probe. The various possible sources of probes will be dealt with below (see pp. 86–88).

Making restriction-site maps

Any moderately long sequence of DNA – say of the order of 10–50 kb – can be uniquely characterized by its distribution of restriction sites, in other words by its *restriction-site map*. The principle is to cleave the DNA with several restriction enzymes both singly and in pairwise combination and to determine the sizes of the fragments generated in each case. It is then generally possible to deduce a unique sequence and spacing of sites. A worked example is set out in Box 3.1. Ambiguities can arise from too-similar sizes of different fragments or from unfavourable modes of interspersion of different kinds of restriction sites; they can generally be resolved by the use of additional enzymes. Sometimes, to be sure of which smaller fragments are derived from which larger ones, it is necessary to carry out sequential rather than simultaneous digestions; a fragment formed by one enzyme can be excised from the electrophoresis gel and cut again on its own by a second enzyme.

Restriction-site mapping is best carried out on a pure preparation of a DNA molecule or fragment, with visualization of the restriction sub-fragments by ethidium staining. However, it is also possible to make a map of a piece of DNA in a complex mixture – even a whole genome – provided one has a probe to show up its various restriction fragments on a Southern blot.

The scale of restriction-site mapping depends on the enzymes used. 'Six-cutting' enzymes, which cleave on average once every few kilobases, are suitable for mapping tracts of DNA in the range 10–50 kb. 'Four-cutters', though they have their special uses, are too small-scale for most purposes. On the other hand, a recently domesticated class of 'eight-cutting' enzymes, that generate fragments of tens or hundreds of kilobases, has become extremely useful for coarser-scale mapping of substantial tracts of chromosomal DNA (see Table 3.1 and Chapter 5, pp. 152–153).

Cloning and cloning vectors

Making recombinant DNA molecules

The modern era of molecular genetics is based on the propagation (*cloning*) of specific pieces of DNA. Molecular cloning both amplifies and purifies specific sequences and so, in principle, enables complex genomes to be analysed in detail piece by piece. It became possible following the discovery of natural DNA molecules with the required properties of autonomous replication and amplification, and the development of methods for splicing alien DNA segments into them as 'passengers' (Fig. 3.2).

End-to-end splicing (*ligation*) of DNA fragments is catalysed by a class of enzymes called *ligases*. The ligase from *Escherichia coli* will only join double-stranded fragments with mutually complementary ('sticky') single-stranded tails, such as those generated by a restriction enzyme making staggered cuts. This limitation is often extremely useful, since it gives the experimenter greater control over the products of ligation. In general, *E. coli* ligase will join only sticky ends generated by the same restriction enzyme. Sometimes, however, it is necessary to join ends of diverse origins, and here the bacteriophage T4 ligase is used since, under appropriate conditions, it will join flush-ended DNA molecules (i.e., with no single-stranded terminal overhangs) in an unspecific way. Fragments with single-stranded 5'-overhangs can be made flush-ended by filling in the ends with DNA polymerase.

Escherichia coli plasmid vectors

Most of the cloning vehicles (*vectors*) in current use were developed from *E. coli* plasmids and, in particular, from plasmid ColE1. Plasmids are double-stranded, usually circular (or, more precisely, closed-loop) DNA molecules, extraneous to the essential bacterial genome and replicating autonomously in the cell. Plasmids, from the bacterium's point of view, can be regarded as molecular parasites or, in some circumstances, as useful partners. They are of numerous different kinds, with a wide range of size and complexity of function. Many of the larger ones (such as F, see p. 37)

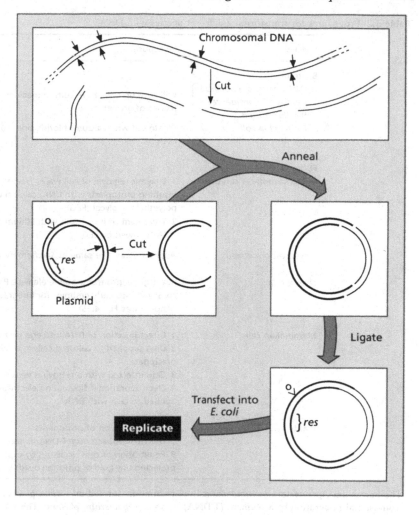

Fig. 3.2 Cloning of DNA restriction fragments by ligation into a plasmid – usually a derivative of ColE1. Chromosomal DNA is cut with a restriction enzyme, generating short single-strand mutually cohesive overhangs at the ends of the fragments. The cloning plasmid is cut in a single site by the same enzyme, converting the closed loop to linear DNA with the same 'sticky' overhangs. A mixture of cut plasmid DNA and chromosome fragments is allowed to anneal, and new circles, many with chromosome fragments joined to plasmid, are stabilized with DNA ligase. The plasmid contains both a replication origin (o) and a gene conferring antibiotic resistance (*res*), so the hybrid circles including plasmid sequence can be selected for and replicated in *Escherichia coli* grown with the antibiotic.

are self-transmissible in the sense that they can be transferred infectively from cell to cell by a mechanism encoded in their own genes. It is common for such plasmids also to carry genes conferring resistance to antibiotics or other toxic agents.

ColE1, which is only 4.2 kb in size, confers two properties on cells carrying it: firstly the production of a *colicin* – that is, a protein that is a highly toxic to most *E. coli* strains – and secondly immunity to the effect of the colicin. It replicates in the bacterial cell to up to 15–20 copies. It is not infectious by itself, but can move from cell to cell in company with any self-transmissible plasmid that happens also to be present. It can also enter *E. coli* cells under the special artificial conditions

devised for uptake of DNA in general, the most crucial factor being the presence of relatively high concentrations of calcium (see Table 3.2).

Like all other plasmids, ColE1 owes its capacity for autonomous replication to the possession of a *replication origin* – a specific DNA sequence at which the DNA polymerase complex can initiate a new round of synthesis. So long as this sequence is present, much of the remainder of the plasmid can be replaced by blocks of DNA from other sources, and the new construct will still be replicated.

In what follows, one widely-used class of ColE1-based vectors will be described; most of its features are shared by the other types in common use. These artificial plasmids, given the general symbol pUC followed by an identifying number, have a

Table 3.2 Procedures for transforming different species with DNA

Species	Procedure
Bacteria	
Diplococcus pneumoniae *Hemophilus influenzae* *Bacillus subtilis*	Will take up DNA from culture medium during certain phases of growth
Escherichia coli	Treatment with calcium chloride prior to exposure to DNA
Fungi	
Saccharomyces cerevisiae	1 Enzymic removal of cell walls. Treatment of resulting protoplasts with DNA, calcium chloride and polyethylene glycol (PEG) 2 Treatment of intact cells with lithium acetate – then DNA followed by PEG
Neurospora crassa	Approximately the same as *Saccharomyces* method 1
Drosophila	Insert gene into transposable element P and inject the construct into early embryos for integration into the genome (see Fig. 4.15)
Mammalian cells	1 Direct injection of DNA into egg pronuclei 2 DNA supplied to cultured cells as a calcium precipitate 3 Cells infected with a retroviral vector 4 Electroporation – fluctuating electrical potential applied to cells with DNA
Plants	1 Electroporation of protoplasts 2 Use of *Agrobacterium* T-DNA as vector* 3 Penetration of cells or tissues by explosively projected tiny gold or tungsten beads coated with DNA†

* The bacterium *Agrobacterium tumefaciens* lives in tumours formed on certain plants as a consequence of transfer and chromosomal integration of a segment (T-DNA) of an *Agrobacterium* plasmic. The T-DNA can be used as a vector for introducing other kinds of foreign DNA into the plant genome.
† This method may also be useful for animal cell transformation.

ColE1 replication origin and two major blocks of non-ColE1 DNA replacing the genes for colicin production and immunity (Fig. 3.3a). One of these blocks contains a gene, derived from another kind of plasmid, conferring resistance to the antibiotic ampicillin. This serves as a selectable marker, enabling the efficient selection of *E. coli* cells that have acquired the vector. The other block is fabricated so as to include both a polycloning (or *polylinker*) sequence for the insertion of DNA fragments for cloning and a system for signalling whether an insertion has been made.

The polylinker, obtained by chemical synthesis, consists, in one design, of 10 different 6-bp restriction sites, mostly end-to-end but partly overlapping. It is part of a sequence of 81 bp comprising 27 amino acid codons which, in turn, is built into the 'upstream' (i.e., translation start) end of the codon sequence of the *E. coli LacZ* gene, which encodes the enzyme β-galactosidase. The additional 27 amino acids do not, in themselves, affect the enzymic function of the gene translation product, but the further insertion of a restriction fragment of a few kilobases into the polylinker site will inevitably do so. The failure of a cell harbouring the vector to produce β-galactosidase is a good

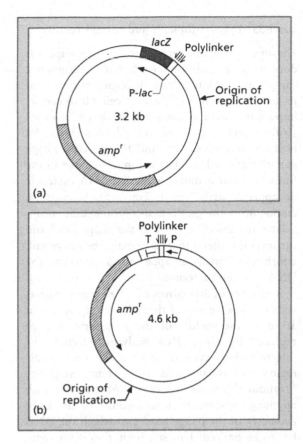

Fig. 3.3 (a) One of the pUC series of cloning vectors. It has a ColE1 origin of replication, a gene (isolated from a drug-resistance plasmid) for ampicillin resistance (direction of transcription indicated) and a *polylinker*, a number of unique restriction sites arranged in close array, just upstream of *lacZ* coding sequence and downstream of the *lacZ* promoter. Insertion of additional DNA into the polylinker prevents transcription of *lacZ*. Bacterial colonies containing the plasmid are selected by ampicillin resistance; those with a plasmid with an insert in the polylinker are distinguished by their lack of β-galactosidase activity (white rather than blue on X-gal medium).

(b) A vector designed for expression of a cloned gene in *Escherichia coli*. It differs from the kind of vector shown in Fig. 3.3a in having the polylinker sandwiched between a strong promoter of transcription P (a hybrid constructed from the *E. coli lac* and *trp* promoters and more effective than either) and a short segment containing sequences (T) that will terminate any transcripts initiated at P. Cloned coding sequences separated from their usual promoters will, if inserted in the polylinker in correct orientation, be transcribed in *E. coli* and, if free of introns (see p. 102) translated as well. Based on Sambrook *et al.* (1989).

indication that an insert is present. (In fact, the *LacZ* segment of pUC vectors codes for only one piece of the β-galactosidase protein, the rest being provided by a complementary gene fragment present in the special *E. coli* host strain, but this complication need not concern us unduly here.)

An important part of the design of the whole vector is that none of the restriction sites in the polylinker is present elsewhere in the vector. Those that were present in the original ColE1 plasmid were removed in the cutting and patching involved in the vector construction. Another result of this restructuring is the breakdown of the natural system of copy-number control. pUC vectors can multiply up to 500 or more copies per cell, a feature that is obviously conducive to bulk preparation of any DNA cloned in them.

The general procedure for securing insertions of alien DNA fragments into plasmid vectors was sketched in Fig. 3.2. It depends upon end-to-end annealing of the mutually-compatible sticky termini of plasmids and DNA fragments cut with the same restriction enzyme, followed by sealing with ligase. The ligated mixture is used to transform calcium-treated *E. coli* cells. If the cloning vector is one of the pUC series, the recipient *E. coli* strain has to carry an appropriate *lac⁻* mutation, and the transformed cells are plated on medium containing ampicillin and a synthetic β-galactoside ('X-gal') that is converted into a blue dye by β-galactosidase. The ampicillin ensures that growing cells have all assimilated the vector DNA and the X-gal discriminates between those growing colonies that are producing the enzyme (blue) and those that are not (white).

Ideally, each white colony will harbour one particular recombinant plasmid carrying a unique DNA fragment in its polylinker site. The cloned insert can easily be recovered from purified vector DNA by cutting it out with the same restriction enzyme that was used to generate it in the first place.

An alternative kind of plasmid vector, designed for expression (i.e., transcription) of the cloned fragment, is shown in Fig. 3.3b. The polylinker is placed just downstream of a strong promoter of transcription and upstream of a transcription terminator sequence. If the cloned fragment includes a complete polypeptide-coding sequence there is a good chance, provided it is inserted in the vector the right way round, that it will be both transcribed and translated. In this case, the polypeptide product may be specifically identifiable in the *E. coli* transformant (see p. 87).

Yeast 2-μm plasmid and shuttle vectors

Closed-circular double-stranded DNA plasmids are not confined to bacteria. A particularly important eukaryotic example is the 2-μm plasmid of *Saccharomyces cerevisiae*. This plasmid, named for the circumferance of its DNA circle (corresponding to approximately 6 kb) is nearly ubiquitous in laboratory strains. It occurs predominantly in the cell nucleus, but can cross at a detectable frequency from one nucleus to another within a binucleate cell.

Like *E. coli* plasmids, the 2-μm plasmid has an autonomous origin of replication and is readily adaptable as a cloning vector. It has several natural restriction sites that can be used to accommodate inserts. Its main usefulness is not as a means of amplification but rather for testing the functions of cloned eukaryotic DNA sequences in a eukaryotic cell environment. Several efficient methods have been devised for introducing free DNA molecules into yeast cells (see Table 3.2) and, indeed, into fungal cells in general, and there is no difficulty about getting yeast plasmids constructed outside the cell propagated within it.

A *shuttle vector* is one that will replicate in either of two different organisms and can therefore be shuttled from one to the other. Yeast–*E. coli* shuttle vectors include the replication origins of both ColE1 and the 2-μm plasmid (Fig. 3.4). Through the use of such hybrid vectors it is possible to use *E. coli* as the host for cloning and amplification of a fragment of eukaryotic DNA (very frequently from yeast itself) and then, without change of vector, to transfer it back to yeast to find out more about its function (see p. 84).

The principle can be extended to shuttling between *E. coli* (or *Saccharomyces*) and mammalian cells, using the replication origin of a mammalian virus, such as *simian virus 40* (*SV40*).

Lambda (λ) bacteriophage and cosmid vectors

λ is an example of a temperate bacteriophage – it does not necessarily kill the infected bacterium but can persist in a latent form through integration of its genome into that of the host cell. However, its importance as a cloning vector depends on its virulent phase, in which it multiplies, lyses the host cell and releases around 100 infective phage particles per cell. λ has a complex genome in the form of a linear double-stranded DNA molecule of approximately 50 kb (Fig. 3.5). It encodes, among other things, the major protein that encapsulates the infective DNA in the phage head and the several proteins that go to make the phage tail, which is essentially an apparatus for injecting the DNA into the bacterium.

A number of derivatives of λ have been used as vectors, making use of the fact that up to about 10 kb in the middle of the λ genome can be replaced by alien DNA without affecting the infection–lysis cycle (Fig. 3.5). This dispensable region can be replaced by a cloned sequence of about the same size (Fig. 3.6), but not with anything very much larger because of the fixed geometry of the phage head into which the DNA has to be packed. This size limitation is the main drawback of λ-based vectors, which otherwise offer important advantages, notably in the convenience of having cloned DNA stably packaged in infective particles. Plasmid-based vectors can take larger inserts, though their efficiency of replication is reduced as their overall size increases. The cosmid series of vectors are combinations of plasmid and λ sequences. They have lost all phage DNA functions except that of being packagable, and they use most of the 50 kb capacity of the phage head for accommodating cloned DNA.

The final stage of replication of λ DNA before it is packaged into phage particles proceeds through what is called the 'rolling circle' mode (Fig. 3.5), the outcome of which is a chain (*concatamer*) of double-stranded genomic copies joined head-to-tail. This chain is cut into single genomes at a particular 12-bp sequence called the *cos* site, which occurs once per genome. Cleavage of this sequence, by staggered single-stranded cuts generating mutually cohesive ('sticky') ends, occurs as part of natural phage assembly from DNA

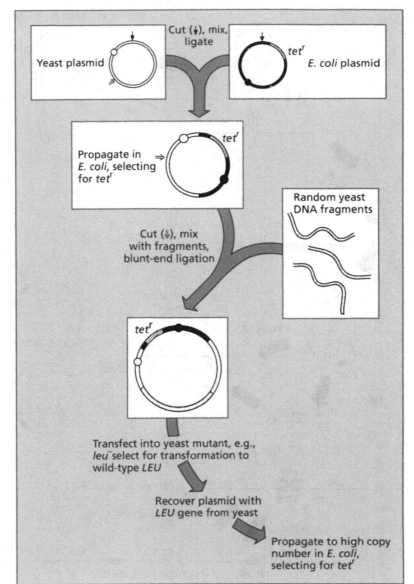

Fig. 3.4 Construction of a shuttle vector from the yeast (*Saccharomyces cerevisiae*) 2-μm plasmid and a ColE1-derived *Escherichia coli* plasmid, and its use in selective cloning of yeast genes. The shuttle vector contains separate initiation sequences (origins) for replication in both yeast (O) and *E. coli* (●), and has a gene for tetracycline resistance (*tet*^{r}) for selection in *E. coli*. Fragments of yeast chromosomal DNA are inserted into the unique restriction site (⇓) and the constructs are introduced into yeast mutant cells that require a certain gene to enable them to grow under restrictive conditions. For example, if selection is made for growth of transformed *leu2* cells on plates lacking leucine, the growing colonies should harbour vectors with *LEU2* inserts. These can then be extracted and propagated to high concentration in *E. coli*. Based on Beggs (1978).

and head and tail proteins. The assembly process ('packaging') can be reproduced in a cell-free system with λ proteins purified from infected bacteria. This *in vitro* system will cut out and package into λ particles any approximately λ-sized tract of DNA provided it is bounded by *cos* sites in the same orientation.

The structure and use of a cosmid vector are explained in Fig. 3.7. It is a closed-loop double-stranded circle of about 5 kb and incorporates; (i) a ColE1 replication origin; (ii) a selective marker (ampicillin resistance); (iii) a polylinker sequence; and (iv) a *cos* sequence. Large DNA fragments can be obtained by incomplete digestion with a suitable restriction enzyme, and ligated with linearized cosmid DNA. The resulting random concatamers are subjected to *in vitro cos*-cutting and packaging. Fragments in the right size range

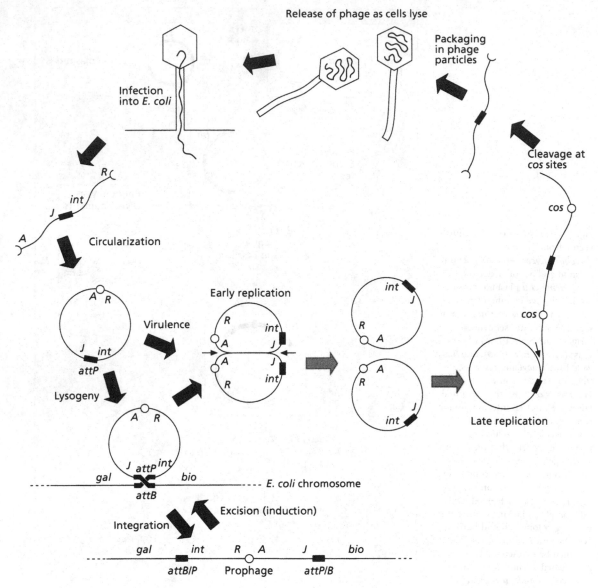

Fig. 3.5 A simplified summary of the alternative survival strategies of bacteriophage *lambda* (λ). The infective particles inject a linear copy of λ DNA which is converted to a closed-loop after entry by virtue of its mutually cohesive 'sticky' single-stranded termini (represented as ( ). There are then two options. The closed-circular DNA may replicate, first bidirectionally to generate more circles, and then with a single replication fork starting at a single-strand nick ('rolling-circle' mode), spooling off a concatemer of λ sequences head-to-tail. The concatemer is cleaved at the cos sequences (shown as ○), producing linear single-copy λ for packaging in phage heads. Lysis of the host cell follows, with release of phage particles. Alternatively, the circularized λ can become integrated as prophage into the λ-attachment site in the *Escherichia coli* chromosome by crossing-over between *attP* and *attB* (shown as black rectangles). Clones of lysogenic cells with integrated λ are immune to lysis and will replicate the latent prophage indefinitely. If immunity breaks down, as it does after ultraviolet irradiation, the prophage is excised and replicated. A few gene markers are indicated: the growth and lysis functions *A, R, J*, and the integration gene *int* in the λ genome; *gal* and *bio* (galactose utilization and biotin synthesis) in the *E. coli* chromosome. Note: all the DNA here is double-stranded but drawn as single lines for simplicity.

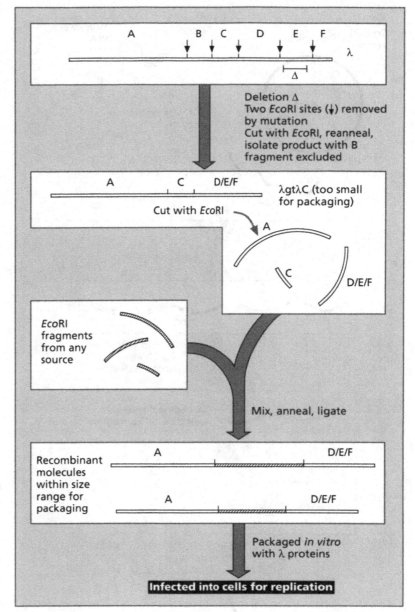

Fig. 3.6 Derivation and use of a cloning vector of the λgt series. Deletion of either *Eco*RI restriction fragment B or C (neither necessary for lytic growth of the phage) leaves the DNA too small for packaging. An *Eco*RI fragment from any other source, within a certain size range (5–12 kb), will, when inserted in place of B or C, enable packaging in infective particles to occur. This packaging is easily performed *in vitro* by mixing the DNA with previously prepared λ head and tail proteins, which self-assemble around any DNA provided it is of suitable size and has *cos* sequences at its ends (cf. Fig. 3.7).

that have become ligated on each side to cosmid elements in the same orientation will end up in λ particles. These contain no λ DNA other than the cleaved *cos* sequence, but they can still inject their contained DNA into *E. coli* cells, where the cosmid clones will be able to replicate as plasmids. The recipient cells can then be selected for the cosmid resistance marker and maintained as a cosmid bank of large cloned fragments of 30–50 kb each.

Yeast artificial chromosomes (YACs)

For molecular mapping of large genomes, dealt with in Chapter 5, it is desirable to be able to increase the size of cloned sequences by yet another

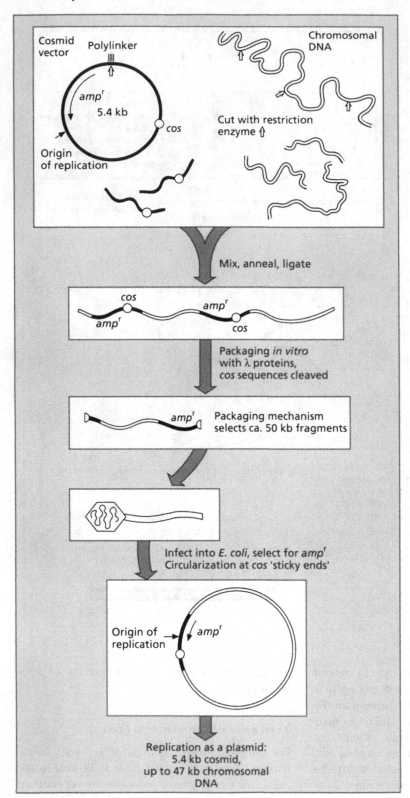

Fig. 3.7 (a) The construction of cosmid libraries of large (30–50 kb) DNA fragments. The cosmid vector (drawn at larger scale) is a plasmid with a ColE1 replication origin, an ampicillin-resistance gene (*amp*^{r}) for selection in *Escherichia coli*, a polylinker, and an uncut *cos* site (○). Chromosomal and cosmid DNA, cut with the same restriction enzyme, are mixed, annealed and ligated, forming concatemers with interspersed chromosomal and cosmid sequences. The λ *in vitro* packaging system cuts the *cos* sequences and packages any approximately λ-sized pieces that have *cos* 'sticky ends' (○). On infection into an *E. coli* cell, the DNA circularizes through its sticky ends and replicates as a plasmid. Based on Sambrook *et al.* (1989).

order of magnitude. The currently favoured vectors for this purpose are YACs, which became possible with the cloning of centromere and telomere sequences. Several DNA fragments including centromeres have been cloned from different *Saccharomyces* chromosomes; they seem to be functionally interchangeable. Telomeres were originally cloned from the ciliated protozoan *Tetrahymena* and the same sequences work in yeast nuclei. The structure and function of telomeres will not be explained in detail here; suffice it to say that they include short repeats that promote replication inwards from the chromosome ends, and are themselves renewable through a special addition mechanism.

The requirements for a YAC are basically simple. There are four essential components: a centromere, telomere sequences at each end, at least one yeast DNA replication origin and two selectable markers. These components can be provided in the form of a kit contained within a special plasmid, an example of which is shown in Fig. 3.8. Cutting the plasmid with restriction enzymes generates the two 'arms' of the YAC, with one 2-μm plasmid replication origin and one centromere between them. Each arm has one of the two selectable markers and a telomere. The very large DNA fragments to be cloned, generated either by very limited mechanical shearing or by rare-cutting restriction enzymes, are joined at random to the YAC arms by flush-end ligation. The ligated mixture is introduced (see Table 3.2) into yeast cells which are then challenged to grow under conditions that require both the YAC marker genes. The colonies that grow harbour YACs, most of which, if the ligation has been performed with an appropriate ratio of arms to fragments, will carry inserts. Fragments of DNA of the order of hundreds of kilobases can be cloned in this way. In fact, YACs with very long inserts behave more like ordinary chromosomes, as judged by their regularity of partition between daughter cells, than those which carry lesser amounts of DNA to pad out the minimal YAC arms.

The success of the YAC strategy is witness to a very remarkable and apparently rather general property of eukaryotic cells – that of being able not only to assimilate large pieces of DNA across

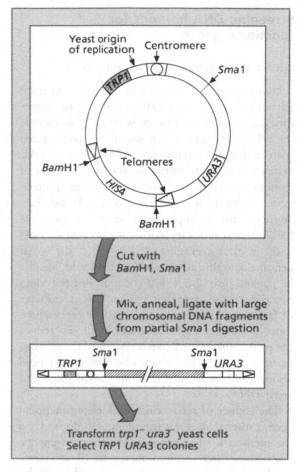

Fig. 3.8 An example of a yeast plasmid constructed for generating artificial chromosomes (YACs) carrying long cloned DNA fragments from any source. It contains a cloned yeast centromere, two telomeric sequences cloned originally from the ciliate *Tetrahymena*, the 2-μm plasmid origin of replication, and the selectable yeast genes indicated. The two artifical chromosome arms are generated by cutting with *Bam*H1 and *Sma*1, and mixed and ligated with an excess of large fragments generated by incomplete digestion with *Sma*1. The ligated mixture is used to transform *trp1 ura3* yeast and selection is made for *TRP1* and *URA3* on unsupplemented medium. The only colonies that grow are those with both artificial chromosome arms, usually with an exotic DNA fragment between them. Based on Burke *et al.* (1987).

the outer cell membrane under certain conditions, but also to admit the alien DNA across the nuclear envelope into the cell nucleus. How this happens is quite unknown; it is difficult to imagine that it has any normal biological significance.

Screening DNA libraries for functional genes

Genomic libraries and cDNA libraries

A collection of DNA fragments cloned from some particular organism is called a *library* or, sometimes, a *bank*. If the library is intended to include the whole genome it will usually be constructed with a cosmid vector or, even better, a YAC system, in order to minimize the large number of clones required. It is possible to make an approximate calculation of the required number. Suppose, for example, one wanted a complete *Saccharomyces* library, consisting of cosmid clones with inserts averaging 45 kb. Since the yeast genome contains about 18 000 kb it could, ideally, be packed into about 400 cosmids. But that does not allow for the fact that, in a random assemblage of fragments, some sequences will be represented more than once and others not at all. To reduce the probability of any particular bit of the genome being missed to less than 1%, one would need about 2000 items in a random cosmid library of yeast DNA.

The chance of recovering a complete functional gene, rather than just a fragment of it, depends on the method used to generate the large fragments. There are three general methods: (i) moderate mechanical shearing; (ii) digestion with a rarely-cutting restriction enzyme; or (iii) incomplete digestion with one of the 'six-cutting' enzymes in common use. If a site within the gene sequence happens to be cut by the restriction enzyme used, one may recover the gene in two or more pieces, so that it cannot be identified by any of the functional tests outlined below. The use of randomly

sheared fragments or incomplete restriction digests may thus be necessary for the cloning of some genes.

If the purpose of a library is to provide a source of DNA sequences that are functional, in the sense of being transcribed, it may be better to construct it, not from genomic fragments, but from messenger RNA (mRNA) sequences reverse-transcribed to make complementary DNA (cDNA). Reverse transcription is catalysed by reverse transcriptase, an enzyme present in the infective particles of *retroviruses*. This class of viruses have RNA as their genetic material. The viral RNA is reverse-transcribed after infection and DNA copies are integrated into the host chromosomes. We return to the significance of retroviruses in eukaryotic genomes in Chapter 5.

A general procedure for cloning double-stranded cDNA is described in Fig. 3.8. The main technical problem to be overcome is the requirement for priming sequences to provide free 3'-hydroxyl termini from which to build not only the first cDNA strand on the mRNA template but also the second cDNA strand complementary to the first. The nearly universal presence of 3' poly-A tails on mRNA molecules is helpful to cDNA preparation in two ways. They first allow the initial purification of mRNA by trapping on, and then elution from, columns containing oligo-dT bound to a cellulose matrix, and then provide a solution to the first-strand priming problem. A linearized cloning vector is annealed to the poly-A through its own artificially-added 5' oligo-dT tail. The other plasmid strand (3' end) can then prime the reverse transcriptase-catalysed reaction to make the first cDNA strand. The RNA is then digested away with a special RNase to make room for the

Fig. 3.9 (*opposite*) One way of constructing a cDNA library. The cloning plasmid contains an ampicillin-resistance gene (*amp*ʳ) and three relevant restriction sites for *Kpn*I, *Hpa*I and *Hin*dIII, shown as K, Hp and H. (a) The plasmid is linearized with *Kpn*I. A 'tail' of oligo-dT is added enzymatically at each 3'-end and then one end is removed by cutting with *Hpa*I. (b) Messenger RNA (mRNA) molecules are annealed through their natural poly-A tails to the oligo-dT-tailed plasmid. cDNA complementary to the mRNA is synthesized by reverse transcriptase, primed by the 3'-end of the oligo-dT. (c) Oligo-dC tails are added to both ends, and then a segment at the plasmid end of the construct is removed by cutting with *Hin*dIII. (d) A plasmid fragment tailed with oligo-dG at one end and with a cut *Hin*dIII site at the other is annealed and ligated to the oligo-dC-tailed end of the cDNA. (e) The construct is circularized through its mutually-cohesive *Hin*dIII-cut ends (see Table 3.1) with DNA ligase. The RNA is nicked and digested with RNaseH (which acts on RNA hybridized to DNA) and replaced by the second cDNA strand, synthesized by DNA polymerase, primed by the 3'-hydroxyl ends of transient RNA fragments. The cDNA, carried in the reconstituted plasmid, can now be propagated in *Escherichia coli*. RNA is shown as ------; plasmid DNA as ———; cDNA as ⌁⌁⌁. After Okayama & Berg (1982).

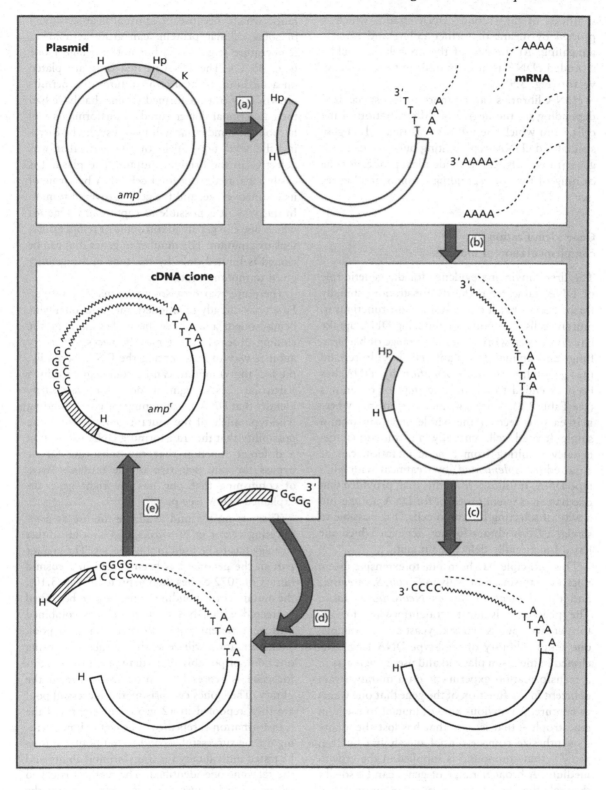

synthesis of the second cDNA strand, which is primed by means of further tailing and primer-annealing. At the end of the procedure, double-stranded cDNA is already built in to the cloning vector (Fig. 3.9).

cDNA libraries can be more or less specialized depending on the degree of differentiation of the cells from which the mRNA was isolated. Highly specialized cDNA, representing only one or a few gene transcripts, can provide ideal probes for the cloning of the corresponding chromosomal genes (see p. 87).

Gene identification by complementation of mutants

The first convincing evidence for the genetic role of DNA came from the demonstration, initially for bacteria, that it could restore lost functions to mutant cells. Methods for inducing DNA uptake have now been worked out for a range of bacteria, fungi and animal and plant cells in culture, and the possibility of *transformation* by DNA has been extended to all major groups of organisms (see Table 3.2). At least in micro-organisms, where it is easy to recover the whole organism from a single treated cell, virtually any functional deficiency resulting from genetic mutation can be repaired (*complemented*) by treatment with wild-type DNA. It follows from this that, provided that one has an efficient protocol for DNA uptake and a way of selecting for cured cells, it is possible to screen a DNA library for any gene for which one has a functionally defective mutant.

This principle has been put to extensive use in micro-organisms, especially in *E. coli, S. cerevisiae* and the filamentous fungus *Neurospora crassa*. The procedure is most straightforward for the former two species. To take yeast as the example, one takes a library of wild-type DNA fragments cloned in the 2-µm plasmid and uses it *en masse* in a transformation experiment on a mutant strain deficient in the function of the gene that one wants to capture. An obvious kind of mutant to use is an auxotroph – that is, one that has lost the ability to synthesize some essential metabolite but can grow if that compound is supplied in the growth medium. A broader range of genes can be sought through the use of temperature-conditional mu-

tants, which, because of unusual heat-sensitivity of some cellular protein, can grow at a lower temperature (e.g., 25 °C) but not at a higher one (e.g., 35 °C). The DNA-treated cells are plated on a medium, or at a temperature, that permits growth only of transformed strains that have had their functional defect cured. Transformants will harbour plasmids that will most usually be carrying the wild type allele of the gene that was defective in the original mutant. The plasmid is easily separated from total cell DNA by virtue of its distinctive size, and the gene recovered from it. In this way, it is possible to capture any gene for which one can get an auxotrophic or temperature-sensitive mutant. The number of genes that can be cloned is limited only by the time and ingenuity given to mutant hunts.

There are two provisos to be made, however. Firstly, as already mentioned, the functional gene being sought may have been disrupted in the cloning process, and it may be necessary to try another way of fragmenting the DNA. Secondly, the fact that a mutant can be complemented by a particular DNA segment does not *necessarily* identify that DNA as containing the corresponding wild-type allele of the mutant gene. There is the possibility that the transforming DNA is boosting a different function that can substitute for or bypass the one defective in the mutant. Ways of confirming that one has the right gene are explained below (see p. 88).

There is no plasmid available for use as a replicating vector in *N. crassa*, and so a less direct strategy has to be used in this species. The greater part of the genome has been cloned in a cosmid library of 3072 clones. In one protocol (Fig. 3.10), the mutant corresponding to the gene to be cloned is treated with DNA from the clones combined into 32 different pools, 96 items in each pool. Transformation will generally be successful using one pool, or possibly more than one if the sought-for gene is represented more than once in the library. The clones comprising the successful pool are then repooled in 12 groups of eight and the transformation is repeated. The eight clones making up the successful second-round pool can then be tested individually for transforming ability and the relevant one identified. The method is called *sib-selection* because what is selected is not the

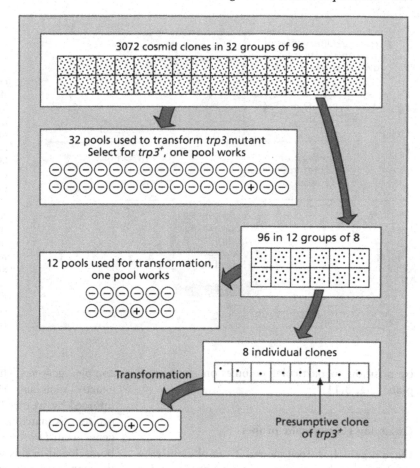

Fig. 3.10 The use of sib-selection for the isolation of gene clones in the fungus *Neurospora crassa*. The example of *trp3* is used here, but numerous other genes have been cloned in the same way. The protocol shown is that of Vollmer & Yanofsky (1986).

actual DNA from the transformants at each stage but rather sister (*sib*) clones from the same pool. Numerous *N. crassa* genes have been recovered as DNA clones in this way.

Complementation as a means of gene identification is much less easily applied to animals or plants. However, some genes from higher eukaryotes have been cloned through their ability to complement yeast mutants. One would only expect this to work with genes whose functions have remained very much the same through the whole of eukaryotic evolution. Some of the genes that meet this requirement function in cell division. A large collection of temperature-conditional mutants have been isolated both in *S. cerevisiae* and in the fission yeast, *Schizosaccharomyces pombe*, and at least one of these (*cdc2* – see p. 185) turned out to be complementable by mammalian cDNA clones.

Screening gene libraries with DNA probes

An alternative strategy, applicable to all organisms, is to screen for genes, or classes of genes, with DNA probes derived from or otherwise related to their RNA transcripts.

The principle of identification here is essentially the same as in hybridizing blots from electrophoresis gels. Assuming that the library is contained in plasmids in cell colonies growing on plates, a sample of cells from the complete pattern of colonies is 'printed' from the plate to a suitable DNA-binding membrane. The membrane is treated with alkali and detergent to lyse the cells and release the DNA. The alkali denatures the DNA to single strands. After treatment to fix the DNA, the membrane is hybridized to a single-stranded radioactively labelled probe and exposed to X-ray film. The spots of radioactivity are related

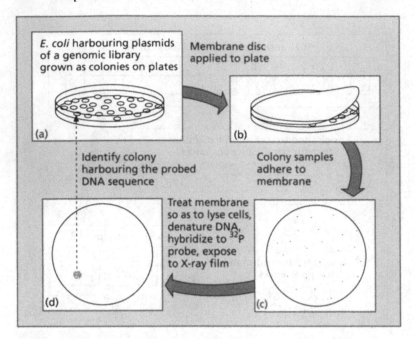

Fig. 3.11 Probing for specific cloned DNA fragments by hybridization of colony prints with radioactive probes. (a) Colonies formed by *Escherichia coli* cells harbouring clones from a genomic library. (b) Transfer of samples of cells from each colony on to a nitrocellulose membrane disc. (c) Lysis of cells on membrane, released DNA (made single-stranded with alkali) bound to the membrane which is then hybridized to radioactive probe and exposed to X-ray film. (d) A radioactive spot recorded on the film indicates the position of a colony harbouring a clone recognized by the probe.

by position to particular colonies on the master plate (Fig. 3.11).

Designing gene-specific probes

Once a gene has been cloned it can obviously be used as a probe for the same gene again, or for another gene with sufficient similarity to it. But how can one devise a probe for a gene never before captured?

Given knowledge of at least a part of the amino acid sequence of a protein encoded by the gene, it is possible to deduce something about the base sequence of the gene transcript and hence of the gene itself. Because of the degeneracy of the amino acid code, the mRNA sequence cannot be deduced exactly from amino acid sequence. However, by choosing a part of the amino acid sequence with minimum coding alternatives, and taking into account the strong bias in favour of certain codons that most organisms display, it is possible to select a panel of candidate coding sequences – say the 20 most probable sequences of 25 bases. With modern automatic DNA-synthesis machines such a panel is not unduly expensive to obtain, and a sequence of 25 is long enough to have a good chance of identifying a unique gene even in a complex genome. The specific signal given by an exactly matching probe will tend to be reinforced by weaker but significant signals due to probes mis-matching in one or a few bases. Many genes from *Drosophila*, mouse and humans have been cloned in this way. It should be noted that this approach can result in isolation of genes for which no viable mutants are known – all that is necessary is accurate information about at least a part of the amino acid sequence of the gene translation product.

There is one favourable situation in which it is possible to devise a unique probe sequence. An example is provided by the *S. cervisiae CYC1* gene, which encodes the major yeast cytochrome c. As we saw in the previous chapter (see Fig. 2.14), many mutations in this gene have been isolated. Some of them consist of single base-pair deletions or insertions and eliminate cytochrome activity by shifting the reading-frame on which decoding of the message depends. In several places along the gene a frameshift mutation was identified that could be reverted to apparent wild type by a compensating insertion or deletion some distance away from the primary one. The effect of the second frameshift in such cases is to restore the correct reading frame, but in between the two

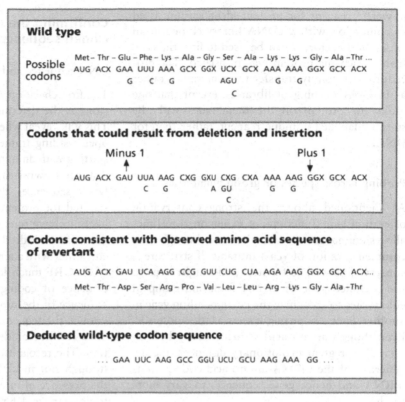

Fig. 3.12 Deducing a DNA sequence of a yeast gene by double-frameshift analysis. The altered amino acid sequence in a revertant, derived from a null *cyc1* mutant can be explained by two successive frameshift mutations, one a single base deletion in codon 3 and the other the insertion of a G in codon 11 or 12. Of the several alternative messenger RNA sequences between codons 4 and 11 consistent with the wild-type amino acid sequence, only one is consistent with the amino acid sequence in the double-frameshift strain. X means any base; vertically aligned bases are alternatives. After Montgomery *et al.* (1978), who used the deduced sequence as a probe for cloning of the gene.

mutually compensating changes the messenger is read out of frame, so that a series of unusual amino acids interrupts the otherwise normal sequence. Knowing the possible codons for each amino acid, one can see that only one particular codon sequence is compatible with the replacement of one amino acid sequence by the other as the result of a frameshift. The argument is summarized in Fig. 3.12. By synthesis of the predicted sequence and its use as a probe, the *CYC1* gene was indeed cloned.

Use of cDNA probes

Occasionally, a few classes of mRNA are so abundant in a differentiated cell type that they can be used almost directly for gene identification. A historically important example is that of globin mRNAs in mammalian (rabbit) reticulocytes (red blood cell precursors). These cells are devoted virtually exclusively to synthesis of globin, the protein of haemoglobin. Adult globin molecules consist of two kinds of polypeptide chain, α and β, each present in two copies. The two corresponding mRNAs were isolated in pure form and used to make cDNA clones which were used as probes in the isolation of the corresponding genes from a rabbit genomic library. The comparison between the genomic and cDNA sequences led to the first identification of eukaryotic introns (see p. 103).

Screening for gene expression

If the protein product of a gene is known (or guessed at) there is another possible avenue to its identification. Provided the protein can be purified, even in small quantities, it is possible to use it to immunize mice. Through what is now a fairly routine procedure, antibody-producing cells can be fused to cancer cells to found immortal hybrid cell clones (*hybridomas*) each of which produces a single *monoclonal antibody* (MAB). A suitably selected MAB is an extremely specific and powerful reagent for protein identification. Used in conjuction with a cDNA library cloned in an expression vector, it can be used to find the gene from which the protein was translated. The procedure uses colony transfer to membranes, rather as in DNA probing of libraries, except that one looks for protein on the membranes with the suitably labelled immune reagent, rather than for DNA.

Probing across species or group boundaries

As mentioned above, the strong conservation of certain gene functions in evolution sometimes allows animal genes to be identified through their complementation of yeast mutants. If structure is conserved as well as function, one might hope to use a cloned gene or cDNA from one group of eukaryotes to recognize the corresponding gene in another. Such a hope might seem a little optimistic. Even though amino acid sequences may be conserved over great evolutionary distances, the degeneracy of the mRNA-amino acid code permits mRNA and hence gene sequences to vary more freely.

Nevertheless, speculative cross-group probing has had some remarkable successes. One of the best examples arose from the study of mammalian oncogenes. These genes, in their normal chromosomally integrated form (cellular oncogenes) have various functions in the control of cell growth and division. They were discovered, in mutant form, integrated into the genomes of oncogenic (cancer-producing) viruses. Cloned DNA from a human member of the *Ras* (rat sarcoma virus) family of oncogenes was used to probe a *Saccharomyces* genomic library, and two *Ras* analogues were found. The question of what functions these genes had in yeast was addressed by engineering mutations in them, as described on p. 90 (Fig. 3.13).

Positional cloning

As chromosome maps become more and more detailed, it is increasingly possible to clone genes on the basis of the mapped positions of the mutations that affect their functions. Consideration of this approach to cloning is deferred until Chapter 6, in the context of the complete mapping of chromosomes at the molecular level.

Confirming the identity of cloned sequences

DNA sequence and open reading frames

The first check on the identity of a putative gene clone is the determination of its DNA sequence. This will reveal whether it includes any significant open reading frame (ORF) – that is, a sequence starting with an initiation codon and ending some distance downstream with a terminating codon. DNA sequences that have not been naturally selected for protein coding function are very unlikely to have runs of codons extending for more than a few hundred nucleotides without a random encounter with a termination codon. The presence of an ORF much longer than that is *prima facie* evidence of coding function. If the amino acid sequence of the supposed gene product is already known, the sequence of codons in the ORF will at once confirm or deny the provisional identification. The recognition of ORFs in genomic clones (though not in cDNAs) may be complicated by the presence of intervening sequences (*introns*), a topic that we defer until the next chapter (see p. 102).

Using the clone to disrupt the corresponding gene

One way of relating a cloned DNA sequence to a gene is to use it to engineer a crippling mutation in the corresponding sequence in the genome. If the gene one is trying to clone has already been identified through a series of mutant alleles, with characteristic phenotypes and defined map position, one can ask whether the engineered mutation is a member of the same series by the criteria of complementation and recombination.

Using a DNA clone to disrupt the corresponding gene is relatively easy in *S. cerevisiae* because of an almost unique feature of DNA-mediated transformation in this species. In other fungi, and also in animals and higher plants, transforming DNA tends to integrate into the chromosomes *ectopically* – that is, within non-homologous sequences at many different loci. Only occasionally does integration occur by homologous recombination between the incoming DNA and the corresponding resident gene. In *Saccharomyces*,

Fig. 3.13 Disruption of the *Saccharomyces cerevisiae RAS* genes by homologous integration of plasmid sequence. (i) Plasmid clone containing a yeast genomic *Eco*RI fragment including a sequence (*RAS2*; stippled) hybridizing to a human *Ras* gene probe; (ii) insertion of *LEU2* into the *Ras*-probed sequence at a site cut by restriction enzyme *Pst1*; (iii) the *Eco*RI fragment with the *LEU2* insert cut out and purified; and (iv) used to transform *leu2* yeast cells. (v) Replacement-type integration confers ability to grow without leucine, but disrupts the *RAS2* gene. Two different *Ras*-related genes, *RAS1* and *RAS2*, were cloned and disrupted in this way. Disruption of either one alone had little or no effect, but the doubly disrupted ascospores from crosses of singly disrupted strains failed to germinate. After Tatchell *et al.* (1984).

(b) The same method modified for gene disruption in mouse cells by inclusion of a counter-selectable marker to eliminate cells transformed by random integration of the DNA construct. The cloned mouse *int-2* gene, which consists of coding exons (heavily stippled boxes) and non-coding introns (open boxes) – see Chapter 4 – was modified: (i) by insertion of a bacterial gene for neomycin-resistance (*neo^r*, hatched box) into one of the exons; and (ii) by replacement of one end by the herpes simplex virus gene encoding the enzyme thymidine kinase (*TK*, stippled box). This modified sequence was used to transform mouse cells which were then cultured on medium containing two drugs: G418, a neomycin analogue, which killed all cells that had not acquired *neo^r*, and gancyclovir, which killed cells that had extra thymidine kinase. To survive, cells had to integrate *neo^r* without *TK*. This can happen by homologous replacement, as shown, but only rarely by non-homologous (ectopic) integration, which usually includes both markers together. After Mansour *et al.* (1988).

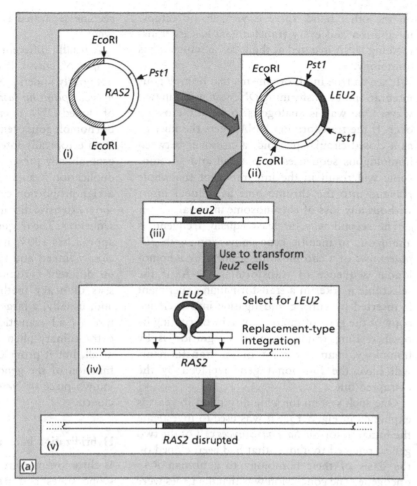

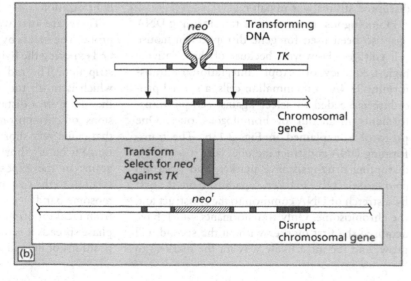

on the other hand, there is virtually no ectopic integration and every transformant has its transforming DNA inserted at the locus to which it has homology.

It seems true in all organisms that homologous integration of transforming DNA can occur in two ways. One way is analogous to meiotic crossing over. If the transforming DNA enters the nucleus as a closed-circular plasmid, a crossover between homologous sequences in plasmid and chromosome will result in the integration of the whole plasmid into the chromosome as a linear insert without any loss of chromosome material.

The second way, at least equally frequent, is analogous to meiotic gene conversion, with *displacement* of a chromosome segment by a homologous sequence of transforming DNA. If the selectable marker in a transformation experiment is inserted by cutting and ligation into a cloned copy of the gene of interest, thereby disrupting its open reading frame, then, if integration is by homology, many or most of the transformants will have the functional gene replaced by the disrupted one.

One such system for gene disruption in yeast is explained in Fig. 3.13a. It was used to investigate the function for the *Saccharomyces* cell of the two genes, referred to above, that had been cloned on the basis of their homology to a human *RAS* oncogene. The conclusion was that the genes were functional duplicates. Disruption of neither gene had any phenotypic effect by itself, but the double mutant obtained from the cross of the two singly-disrupted strains was inviable.

Homologous integration of transforming DNA has also been used for gene disruption in mouse cell cultures. However, because of the relatively high frequency of ectopic integration of transforming DNA in mammalian cells, a special procedure is needed to select against ectopic transformants in favour of homologous ones. One protocol is explained in Fig. 3.13b. The transforming DNA construct includes both a selectable disrupting drug-resistance marker, and a second *counter*-selectable marker, separated from the first by a stretch of DNA common to the construct and the chromosome. Only transformants which have acquired the first marker without the second will grow, and the most likely way for the markers to

become separated is by a homologous exchange event.

A totally different method is applicable to the fungus *N. crassa*. It depends on a highly efficient but poorly understood phenomenon called *repeat-induced point mutation* (RIP). Ectopic integration of cloned DNA is easily obtained in this species but homologous events are generally harder to find. If an ectopically integrated sequence duplicates a gene already present at its normal locus, and the duplication strain is crossed with any other strain, a high proportion of the resulting meiotic tetrads show extensive disruption of both of the duplicated sequences. The frequency of these aberrant tetrads approaches 100% if the duplicated sequences are closely linked and typically 30–50% if they are on different chromosomes. The disruption consists of heavy methylation of cytosine residues and, usually, a large number of base-pair substitutions, all transitions of G–C to A–T. This extraordinary phenomenon is hardly understood at all, but it provides a means of destroying the function of the genomic counterpart of any unknown piece of *Neurospora* DNA that has been cloned.

Hybridization back to chromosomes

If chromosomes are spread and fixed on microscope slides and treated to make their constituent DNA single stranded, a DNA probe will bind to any chromosomal locus to which it has sufficiently close homology. This is called *in situ* hybridization.

There are two ways of visualizing an *in situ* probe. The first is by labelling it with radioactivity and exposing the hybridized slide to radiosensitive strip film. The radioisotope required is not ^{32}P, which is much too energetic and would blacken the film over a distance far exceeding the dimensions of chromosomes, but rather tritium (^{3}H) that emits very short-range radiation. Even so the range is barely short enough. The scatter of silver grains in the exposed film is not too great in relation to the size of a *Drosophila* polytene chromosome band (see Fig. 5.8), but the method is far from ideal for ordinary chromosomes. With metaphase spreads it is necessary to count and note the positions of individual silver grains in a sample of a hundred or so preparations so as to establish a significant tendency for grains to occur over a particular rather broadly defined chromosome segment.

Much better definition can be achieved by a method sometimes abbreviated to FISH – *fluorescent in situ hybridization*. Briefly, the DNA probe is ligated to an antigenic protein that will bind tightly to a specific antibody. The antibody itself is coupled to a dye that fluoresces intensely in ultraviolet light. The chromosomal DNA is hybridized to the antigen-bound probe, which is then detected in turn by the fluorescent antibody. The sites of hybridization can be defined with a precision limited only by the resolving power of the microscope. Through the use of different dyes fluorescing in different colours, several gene loci can be defined independently in the same chromosome preparation (Ried *et al.*, 1992).

Simultaneous mutagenesis and gene tagging

The principle

So far in this chapter we have considered ways of isolating the DNA of a gene already identified by mutation, and of targeting mutations to genes or parts of genes isolated in DNA clones. These two steps of analysis can sometimes be brought together through the exploitation of naturally occurring *mobile genetic elements* that can insert themselves into or adjacent to virtually any gene. Disruption by such elements is responsible for a high proportion of 'spontaneous' gene mutations in several of the best-investigated organisms, notably maize, yeast and *Drosophila*. In disrupting the gene, they also introduce into it a known sequence that can be probed for.

Transposon tags

Mobile DNA elements of various kinds are major components of the genomes of most eukaryotes. We consider their significance as components of whole genomes in Chapter 5. In the present context we need consider only two classes, retrovirus genomes and transposable DNA sequences of the kind characterized as controlling elements by Barbara McClintock in maize (see Fedoroff, 1989, for a review).

Retroviruses contain RNA, not DNA, as their infectious genetic material. After entry into the host cell, the virus RNA is reverse-transcribed into DNA through the activity of the virus-encoded enzyme reverse transcriptase. The reverse transcript is then made double-stranded and inserted into some site in the host genome as a *provirus*. From its chromosomal retreat it can be transcribed into virus-specific RNA for translation into reverse transcriptase and the various other particle proteins and inclusion into a new generation of virus particles. The insertion of provirus is not completely at random since some kinds of sequence are better targets for insertion than others, but it can certainly occur at a very large number of different sites. It has been shown that several different long-standing mouse mutants owe their origin to disruptive insertion of provirus.

With the development of methods for raising whole mice from cultured embryonic stem cells, it is becoming possible to generate new kinds of mutant mice by retroviral infection of cells in culture. The infective agents used are non-pathogenic retrovirus derivatives with a selectable marker replacing some of the essential virus genes. If a new and interesting mutant is obtained, the DNA of the disrupted gene can be cloned by virtue of its propinquity to the provirus, for which a specific probe will be available. Fragments of the disrupted gene can then be used as probes for the wild-type gene.

Different strains of maize (*Zea mays*) harbour movable elements of several different classes in their chromosomes. They can transpose from one chromosomal locus to another by excision and reinsertion without any RNA intermediate. Very similar elements (*transposons*) have been found in the snapdragon (*Antirrhinum majus*). They usually occur in multiple copies, some apparently functionally complete with the ability to transpose autonomously, and the others with internal deletions or rearrangements and able to transpose only if a fully functional element of the same class is present in the same nucleus. Transposability is dependent on the presence of repeated sequences in mutually inverted orientation at the ends of the transposons. Transposition depends on the presence of an intact open reading frame in at least one copy. Presumably, the ORF encodes an enzyme that is instrumental in the excision–reinsertion process, and the significance of the inverted repeats is that they provide the same specifically cleavable sequence at each end.

The maize transposable elements were discovered by McClintock through their spectacular effects as determinants of mutability. Briefly, she showed that many maize mutations (mostly affecting seed pigmentation) were due to transposon insertion and that their high frequencies of reversion, giving spotted or sectored phenotypes, were due to transposon excision. Sometimes a transposon excised from one locus could be shown to have reinserted at another.

Several transposons from both maize and antirrhinum have now been cloned, and there are a number of examples of their use in gene tagging. We shall consider in detail one example from antirrhinum, since it illustrates several of the recognition techniques reviewed in this chapter. The analysis included the following steps (Fig. 3.14).

1 A recessive mutant with snow-white flowers (*nivea*) was shown to produce no flower pigments of any kind. It was surmised that the wild-type *Niv* gene encoded the enzyme chalcone synthase (CS), chalcone being the common precursor of all the missing pigments.

2 One mutant allele, *niv-recurrens* (*niv*[rec]), gave abundant coloured spots and stripes on a white background, a phenotype thought likely to be due to the transposition into the *Niv* locus of a transposon from a mutable allele, *pal*[rec], of another gene encoding the enzyme for the last step in the pigment synthesis *Pallida* (*Pal*).

3 A cDNA probe for the CS gene was already available from another plant species. The synthesis of CS in cultures of parsley cells had been found to be strongly stimulated by light. A cDNA library, obtained from mRNA of illuminated parsley cells and made with an expression vector, was screened for clones that hybridized only with light-induced mRNA. Clones meeting this requirement were further selected for ability to express CS, which was recognized through use of a specific anti-CS antibody.

4 A cDNA clone of the parsley CS gene was used as probe for cloning the homologous gene from the wild-type antirrhinum genome. It was sequenced to reveal an open reading frame of convincing length.

5 The wild-type (*Niv*) antirrhinum clone was used in turn for probing and cloning DNA from *niv*[rec]. Comparison of restriction-site maps of the *Niv* and *niv*[rec] clones revealed a 3.5-kb sequence (Tam3) inserted, not in the ORF itself, but at a closely placed upstream site, apparently in the promoter region.

6 The Tam3 element was excised from the *niv*[rec] clone and recloned on its own. It was used as a probe in the cloning of *pal*[rec], which then provided a probe for wild-type *Pal* (Fig. 3.14). It has since been used for cloning a variety of genes that display mutability as the result of Tam3 insertion and thus bear the Tam3 tag. Some of these will be mentioned in Chapter 7 in connection with genetic control of flower development.

Conclusion

We saw in the last chapter how genes were shown by genetic methods to have linear structure with coding functions, and in this chapter how DNA fragments with particular functions can be cloned and identified from the mass of genomic DNA. The gene turns out, however, to have unexpected complexities. The next chapter will show how a combination of genetic and molecular methods have been used to extend and refine our idea of the gene.

References

Beggs, J.D. (1978) Transformation of yeast by a replicating hybrid plasmid. *Nature*, **275**, 104–9.

Burke, D.T., Clarke, G.F. & Olson, M.V. (1987) Cloning of large segments of exogenous DNA into yeast by means of artificial chromosome vectors. *Science*, **236**, 806–12.

Fedoroff, N.V. (1989) Maize transposable elements. In Berg, D.E. & Howe, M.M. (eds) *Mobile DNA*, pp. 375–412. American Society for Microbiology, Washington, DC.

Jeffreys, A.J. & Flavell, R.A. (1977) The rabbit β-globin gene contains a large insert in the coding sequence. *Cell*, **12**, 1092–108.

Mansour, S.L., Thomas, K.R. & Capecchi, M.R. (1988) Disruption of the proto-oncogene *int-2* in mouse embryo-

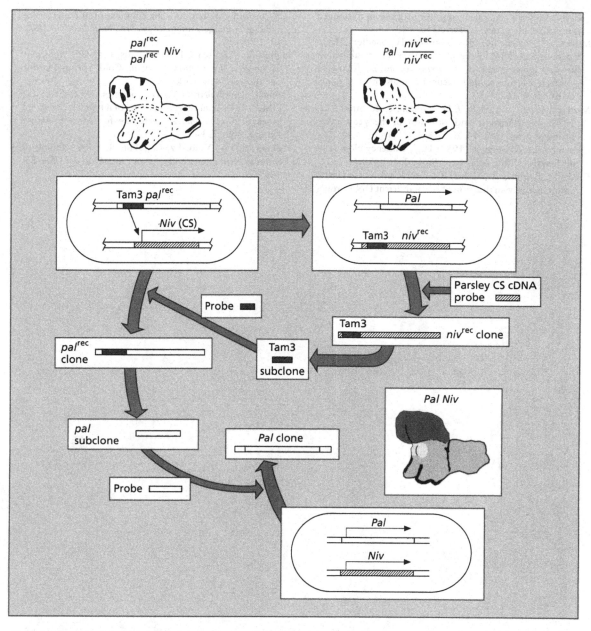

Fig. 3.14 The use of transposon-tagging for cloning the *Pal* gene of the garden snapdragon, *Antirrhinum majus*. *Nivea* (*Niv*) and *Pallida* (*Pal*) are flower-colour genes encoding different enzymes of the pigment (anthocyanin)-synthesizing pathway. The null *niv* allele gives snow-white flowers when homozygous; null *pal* when homozygous gives ivory flowers, yellow (shown as stippling) on the lower lip. The investigation started with the unstable allele *pal*rec (red flakes on an ivory background) and the likelihood that *niv*/*niv* was deficient in chalcone synthase (CS), the enzyme catalysing the first step of pigment synthesis. A *niv*rec allele (red flakes on a snow-white background) was obtained by breeding from a strain in which *Niv* was present together with *pal*rec. The hypothesis was that *Niv* had acquired a transposable element by transfer from *pal*rec. CS messenger RNA, originally from parsley, was used to make a copy DNA (cDNA) probe for cloning *Niv*. The CS cDNA was then used to probe for and clone *niv*rec, which turned out to be *Niv* with a Tam3 insertion. A Tam3 subclone from the cloned *niv*rec was then used to probe a DNA library from *pal*rec. A Tam3-containing fragment uniquely present in *pal*rec was cloned and shown to contain Tam3 inserted into *Pal*. A *Pal* subclone could then be used as a probe for cloning wild-type *Pal*. After Martin *et al.* (1985).

derived cells: a general strategy for targeting mutations to non-selectable genes. *Nature*, 336, 348–52.

Martin, C., Carpenter, R., Sommer, H., Saedler, H. & Coen, E.S. (1985) Molecular analysis of instability in flower pigmentation in *Antirrhinum majus*, following isolation of the *pallida* locus by transposon tagging. *EMBO J*, 4, 1625–30.

Montgomery, D.L., Hall, B.D., Gillam, S. & Smith, M. (1978) Identification and isolation of the yeast cytochrome c gene. *Cell*, 14, 673–80.

Okayama, H. & Berg, P. (1982) High efficiency cloning of full length cDNA. *Mol. Cell. Biol.*, 2, 161–70.

Ried, T., Baldini, A., Rand, T.C. & Ward, D.C. (1992) Simultaneous visualization of seven different DNA probes by *in situ* hybridisation using combinatorial fluorescence and digital imaging microscopy. *Proc Natl Acad Sci USA*, 89, 1388–92.

Sambrook, J., Fritsch, E.F. & Maniatis, T. (1989) *Molecular Cloning: a Laboratory Manual*. Cold Spring Harbor Laboratory, New York.

Tatchell, K., Chaleff, D.T., DeFeo-Jones, D. & Scolnick, E.M. (1984) Requirement of either of a pair of *ras*-related genes of *Saccharomyces cerevisiae* for spore viability. *Nature*, 309, 523–7.

Vollmer, S.J. & Yanofsky, C. (1986) Efficient cloning of genes in *Neurospora crassa*. *Proc Natl Acad Sci USA*, 83, 4867–73.

4

THE EVOLVING CONCEPT OF THE GENE

single gene is functionally indivisible in that no part of it can function when separated from the rest. It follows from the first assumption that, if *a* and *b* are in different genes, the *cis* and *trans* double heterozygotes, *ab*/++ and *a*+/+*b*, must be phenotypically identical – approximating to wild type if *a* and *b* are recessive. It follows from the second that, if *a* and *b* are in the same gene, the *trans* combination *a*+/+*b* will be no more like wild type than the less extreme of the individual mutants. This is the *complementation test*, and it is the criterion generally used in practice.

In terms of molecular biology, the *cis*-principle applies at several different levels. Protein-encoding genes are *cis*-units of translation and all active genes are *cis*-units of transcription, though these often turn out to be far more complex and versatile than was initially imagined. Beyond the transcription unit (sometimes far beyond) one finds *cis*-units of control of transcription. But at each level the *cis*-principle encounters intriguing exceptions or apparent exceptions that we shall deal with as we go along.

The gene as a unit of translation

Single genes – single enzymic functions

There are many examples, especially in microorganisms (e.g., Fig. 2.8), of genes in which the genetic map of mutational sites can be correlated with the sequence of altered amino acid residues in the polypeptide chain of a specific protein, usually an enzyme. Even before genes could be cloned and sequenced, such examples provided ample evidence that the amino acid sequences of

The gene as a *cis*-acting unit

We saw in Chapter 2 how Seymour Benzer, on the basis of complementation and recombination tests on bacteriophage T4 *rII* mutants, was able to make a clear distinction between the genetic unit of function, the *cistron*, and the minimum unit of genetic recombination. The word *gene*, which was used to refer to both kinds of unit before the distinction between them was recognized, was too convenient and well established to abandon, and it came to be used as more or less synonymous with cistron – a functional unit subdivisible by recombination into some hundreds or thousands of linearly arranged individually mutable sites.

The *cis*-principle, as we may call it, promises to provide clear criteria for deciding whether two linked mutations *a* and *b* fall into the same or into different genes. The two assumptions are that different genes function quite independently of their spatial relationship to each other, and that a proteins were encoded in the linear structures of genes. Now, with the development of DNA cloning and sequencing technology, this point is far more quickly demonstrated.

All mutants with altered structure of a particular protein might be expected to fall into a single unequivocal complementation group or cistron except in the case where the protein consists of two or more different polypeptide chains, when the expectation would be two or more quite distinct complementation groups. This expectation is fulfilled most of the time but exceptions are rather common, particularly in yeasts and filamentous fungi where large numbers of mutants affecting the same enzyme protein can often be isolated.

It is common to find a series of point mutants that are defective in the same enzyme and fail to complement one another in *trans*. In such a case one has no hesitation in concluding that the mutants are *allelic* – that is, all due to mutation in the same gene. However, it frequently happens that, as more mutants are looked at, occasional pairs do show a degree of complementation, sometimes to restore something approximating to normal phenotype. This is called *allelic complementation*, an apparent contradiction in terms.

Patterns of allelic complementation can be summarized in *complementation maps* in which mutants are represented by bars – non-overlapping or overlapping depending on whether they show complementation or not. The earliest complementation maps were one dimensional (e.g., Fig 4.1), encouraging the idea that they might reflect in some way the linear structure of the gene. But as more extended series of mutant alleles were analysed, the maps tended to become complex, with branches and loops.

Although allelic complementation may give a wild phenotype at the level of growth or morphology, biochemical analysis of the enzyme product generally reveals relatively low activity and/or reduced stability. This accords with the general explanation for allelic complementation, which is that it results, not in a truly wild-type product but rather in a hybrid protein formed from different mutant derivatives of the same polypeptide chain. There are two demonstrated ways in which this can happen.

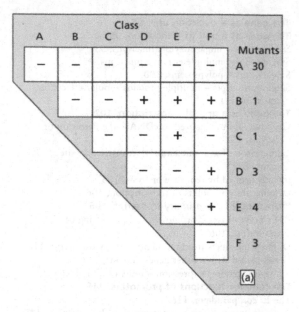

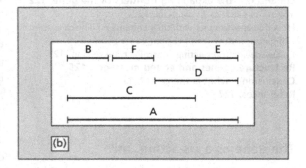

Fig. 4.1 (a) Complementation relationships of *arg-1* mutants of *Neurospora crassa*, and (b) the complementation map derived from them. Forty-two mutants were divided into six classes A–F, with members of the same class showing the same complementation pattern. Mutants were co-inoculated in all pairwise combinations to form heterokaryons. Growth of heterokaryons in the absence of arginine shown as +; absence of growth as −. After Catcheside & Overton (1959).

The first can apply if, as is very commonly the case, the normal enzyme is a dimer or oligomer composed of two or several polypeptide chains of the same kind. Then, when two different mutant polypeptides are synthesized in the same cell, a certain proportion of mixed enzyme molecules will result. To take the simplest case, if equal quantitites of two mutant peptides form dimers at random, the result will be a 50% yield of mixed dimers together with 25% each of the two pure mutant dimers. In dimeric or oligomeric proteins the polypeptide subunits are packed together in a symmetrical fashion with the shape (conformation) of each constrained by its neighbours. Sometimes the reason why a particular mutant polypeptide forms no active enzyme is that, by itself, it tends to lapse into a conformation that is incompatible with its potential catalytic activity. However, when packed into a dimer or oligomer with another mutant polypeptide, inactive for a different reason, it may be forced into its active conformation. The idea is explained for a hypothetical tetramer in Fig. 4.2a.

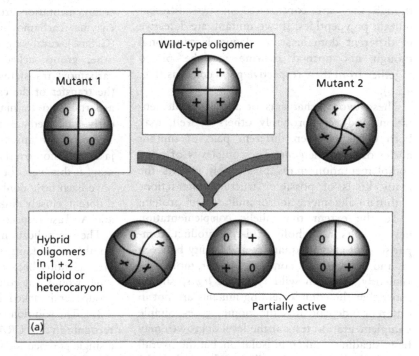

Fig. 4.2 Two general mechanisms of complementation between protein products of mutant alleles. (a) Conformational correction within a dimeric or oligomeric protein (shown here as a tetramer). One monomer is inactive because of loss of an essential amino acid side-chain (O), and the second inactive because of a faulty conformation (×). When the two are packed together, the first may induce the correct conformation (+) of the second, releasing its potential activity (Fincham, 1966).

(b) Reconstruction of a functional protein by piecing together separated domains of what is normally a single polypeptide chain. The shapes attributed to the domains in the diagram are imaginary, to illustrate the principle of specific mutual fit between different domains of a globular protein. Complementation of this kind has been observed, for example, between mutants of *Escherichia coli lacZ* (encoding β-galactosidase) with deletions or chain termination causing non-overlapping deficiencies (Goldberg, 1969).

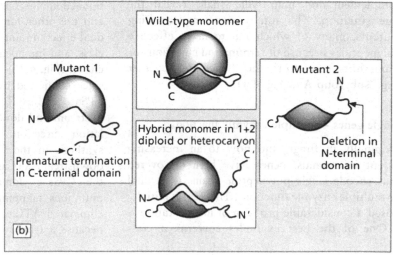

The second mechanism does not depend on dimer or oligomer structure but rather on the common organization of proteins into *domains* – compact globular structures connected by relatively unstructured stretches of polypeptide. The packing together of the different domains is stabilized by their precise complementary fit and a variety of non-covalent bonds, especially by hydrophobic (i.e., water-repellent) interactions. It is often perfectly possible for two mutually fitting domains to pack together in a stable way even though they are contributed by two different mutant polypeptides. If two mutants are defective in different domains, an adequately functional though not normal enzyme may be pieced together from their respective good domains (Fig. 4.2b).

These two mechanisms of allelic complementation, and very probably others as well, may both occur between different pairs of mutant alleles of the same gene. The complexity of some complementation maps presumably reflects the many kinds of possible structural interactions within an oligomeric and/or multidomain protein.

To the extent that allelic complementation never results in a wholly wild-type product, comparisons of enzyme quantity and quality between *cis* and *trans* mutant combinations (or, more conveniently, between wild type and *trans*) should show that the complementing mutants are not in different independently functioning genes. Simple complementation tests at the level of growth may be misleading if taken in isolation, but the overall complementation map will usually reveal the true situation. The totally non-complementing mutants, many of which fail to make effective polypeptides because of termination mutations or frameshifts, serve to tie the allelic series together (e.g., subgroup A in Fig. 4.1).

Single genes – multiple enzymic functions

Originally in fungi, and now to an increasing extent in animals, genes have been discovered that encode single polypeptide chains with dual or multiple enzymic functions. These cases initially posed a considerable problem of interpretation.

One of the best examples concerns a class of auxotrophic mutants, found in three different fungi – *Saccharomyces cerevisiae*, *Neurospora crassa* and the mushroom-like Basidiomycete, *Coprinus radiatus* – that would only grow if supplied with a pyrimidine or pyrimidine nucleoside in the growth medium. Essentially the same situation was described in all three species; here we will refer to the yeast mutants, called *ura2*.

All the *ura2* mutants were mapped genetically to a linear array of closely linked sites. Most of them could be assigned to one or other of two complementation groups. One group lacked the enzyme carbamoyl-phosphate (CP) synthetase (CPSase), catalysing the synthesis of CP, and the other group lacked the enzyme aspartic acid-carbamoyl transferase (ACTase) responsible for the transfer of the carbamoyl group from CP to aspartic acid to make carbamoyl-aspartate, an early intermediate in pyrimidine biosynthesis. These two groups of mutations fell into two adjacent but non-overlapping sections of the genetic map. If this were the whole situation there would have been little doubt that *URA2* was really two different closely linked genes, encoding CPSase and ACTase respectively.

The complication was that a considerable number of mutants, several of which were clearly not deletions because they could revert to wild type, would not complement either of the main groups and lacked both enzyme activities (Fig. 4.3a). The solution of the problem came with the recognition that *URA2* was a single gene encoding a single polypeptide with two independently functional domains, one responsible for CPSase activity and the other for ACTase activity. Most of the dual-effect mutants were of the chain-termination class and mapped within the CPSase-negative cluster (Fig. 4.3b). This prompted the hypothesis, since confirmed by molecular analysis, that the CPSase and ACTase domains were respectively upstream and downstream with respect to translation direction. Termination of polypeptide synthesis in the CPSase domain eliminates the downstream ATCase domain as well as disrupting the CPSase domain. Some chain-terminating mutations mapping within the ATCase domain eliminated ATCase but not CPSase, presumably because a truncated polypeptide chain with the

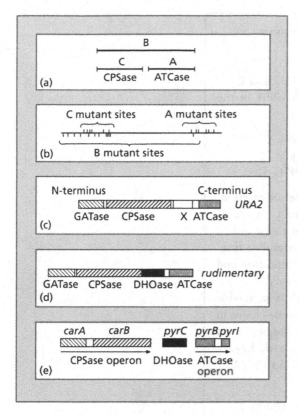

Fig. 4.3 Organization of the genes responsible for the first four enzyme activities in the pathway of pyrimidine biosynthesis. (a) Complementation map of *Saccharomyces ura2* mutants. Class A mutants lack ATCase, class C lack CPSase, and class C lack both enzyme activities. GATase was not distinguished from CPSase at the time this work was done. (b) Genetic map of *URA2*, showing the positions within the map of A, B and C mutant sites. After Denis-Duphil & Kaplan (1976). (c) Structure of the cloned *URA2* gene, showing the regions of the polypeptide product corresponding to three enzyme activities, determined by sequence similarity to the separate *Escherichia coli* genes shown in (e). The apparently functionless region marked X has weak but significant similarities to DHOase genes. After Souciet *et al.* (1989). (d) Structure of the transcript of the *Drosophila melanogaster rudimentary* gene, encoding a polypeptide chain with four enzyme activities. After Freund *et al.* (1986). (e) *E. coli*: five separate genes, two pairs organized as single units of transcription (operons), at three widely separated locations. The *car* operon includes *carA* and *carB*, encoding GATase and CPSase, respectively. The *pyrBI* operon encodes two subunits of ATCase; the *pyrI* subunit is responsible for down-regulation of enzyme activity by the pyrimidine end-product (Neuhard & Nygaard, 1987).

upstream domain intact can still exhibit CPSase activity. It is worth noting that elimination of downstream but not upstream activity could result from a frameshift as well as from a chain-termination mutation.

A note on the genetic identification of chain-terminating mutants is in order here. Such mutants were first recognized in *Escherichia coli* and yeast through their property of being phenotypically *suppressible* by certain kinds of mutational changes in the anticodon sequences of transfer RNA (tRNA) molecules. This kind of suppressibility is a convenient criterion for the identification of chain-termination mutants in micro-organisms. Some examples of how this kind of suppression works are explained in Box 4.1.

In *Saccharomyces* and other fungi the enzyme activities responsible for the first two steps of pyrimidine synthesis are encoded by a single gene. The first enzyme activity, CPSase, has now been shown to include an additional component, glutamine amidotransferase (GATase), which mobilizes the nitrogen for CP synthesis. Thus the *URA2* protein product is trifunctional (Fig. 4.3c). The enzyme for the next step of pyrimidine synthesis, the conversion of carbamyl aspartate to dihydro-orotic acid (DHOase) is in yeast the responsibility of a separate gene. In *Drosophila melanogaster* (Fig. 4.3d), the four enzyme activities GATase, CPSase, ATCase and DHOase are all combined in a single polypeptide chain encoded by the gene *rudimentary* (*r*) (named for the effect of its mutations on the wing). A similar gene occurs in mammals. The pattern of complementation shown by *Drosophila r* mutants resembles that for *URA2* of yeast in that the different enzyme functions can be lost by mutation either individually or together.

From the point of view of molecular evolution, a topic generally neglected in this book, it is interesting to note that in the bacterium *E. coli* these enzyme activities are all due to separate polypeptide chains encoded by different genes (Fig. 4.3e). The four domains of the single *Drosophila r* gene resemble the separate *E. coli* genes not only in function but, to a very significant extent, in detailed sequence also.

The situation of one gene–one polypeptide

Box 4.1 Molecular explanation of mutations suppressing chain-termination

· ·

In *Saccharomyces cerevisiae* the tyrosine codons UAU and UAC are normally recognized by tRNA molecules with GΨA in the anticodon loop:

tyr-mRNA codon	5'-U-A-U-3'	5'-U-A-C-3'
	‖ ‖ ‖ or	‖ ‖ ‖‖
tyr-tRNA anticodon	3'-A-Ψ-G-5'	3'-A-Ψ-G-5'

Note that codon and anticodon pair in anti-parallel orientation, and that anticodon 5'-G will pair with either U or C in the 3' codon position. Ψ stands for pseudouridine, with the base linked to deoxyribose through C1 rather than N2 of the pyrimidine ring; it is derived from uridine by post-transcriptional modification and the base-pairing properties of the uracil are not changed.

Two classes of suppressor mutations, *Su-a* and *Su-o*, enable tyrosine residues to be inserted into growing polypeptide chains in response to the chain-termination codons UAG ('amber') and UAA ('ochre'). The amber- and ochre-suppressors each have a single base-change in the anticodon of a tyrosine-tRNA to CΨA or UΨA. Thus:

amber codon	5'-U-A-G-3'	ochre codon	5'-U-A-A-3'
	‖ ‖ ‖‖		‖ ‖ ‖
Su-a tyr-tRNA	3'-A-Ψ-C-5'	*Su-o* tyr-tRNA	3'-A-Ψ-U-5'

Another class of yeast amber suppressor mutants insert serine in response to the amber codon, through a ser-tRNA anticodon change from CGA to CUA:

one Serine codon	5'-U-C-G-3'
	‖ ‖‖ ‖‖
anticodon of one ser-tRNA	3'-A-G-C-5'
Amber codon	5'-U-A-G-3'
	‖ ‖ ‖‖
Su-a ser-tRNA anticodon	3'-A-U-C-5'

Note. There are multiple genes for all tRNAs, so these suppressor mutants still have normal tRNAs for translation of Tyr and Ser codons.

---

chain—several enzyme activities turns out to be rather common, especially in catalysis of biosynthetic pathways. A second example from *Drosophila* is cited below in a different connection (see pp. 113–114).

Single genes – polyproteins

In the examples just considered, the gene products, while enzymically complex, are still single polypeptide chains. Proteins consisting of discrete domains with separate enzyme activities have often posed a problem to protein chemists. They tend to be susceptible to cleavage by proteolytic enzymes in the inter-domain regions during purification from crude cell extracts. It quite often happened that the purified preparations consisted of separated domains, which were consequently misinterpreted as independent enzymes.

Such inter-domain cleavage can occur, in fact, as a normal *in vivo* event. The best-investigated examples come from *retroviruses*, which translate a long primary transcript into a long single polypeptide chain, which is subsequently cleaved into separate proteins to serve different purposes. There is the additional complication that the long transcript can be processed in different ways to give messenger RNAs (mRNAs) representing different sets of genes. The retrovirus situation, which strains the 'gene' terminology to the limit, is summarized later in this chapter (see Fig. 4.11).

Single transcripts – multiple proteins – polar effects on translation

In *E. coli* and bacteria generally, polypeptide-encoding genes of related function are often

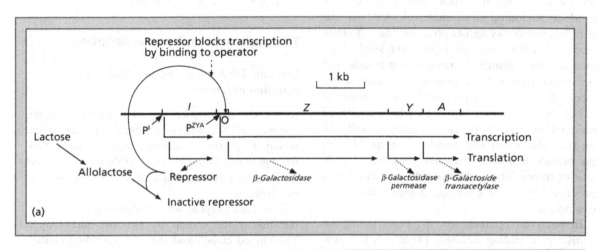

(a)

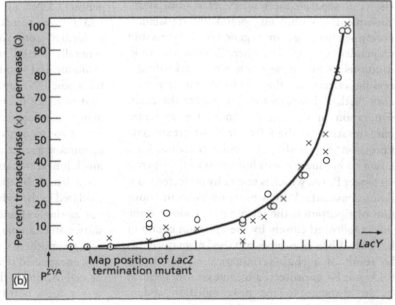

(b)

Fig. 4.4 (a) The operon organization and mode of regulation of the *lac* genes of *Escherichia coli*. O and P are operator and promoter, respectively. (b) Polar effects of termination mutants within *lacZ* (β-galactosidase) on expression of *lacY* (β-galactoside permease) and *lacA* (galactoside transacetylase), expressed as per cent of wild-type level. P^I, P^{ZYA} are promoter sequences. Data from Newton *et al.* (1965).

organized in *operons* − that is, in single units of transcription containing two or more open reading frames. The paradigm example is the *E. coli lac* operon (Fig. 4.4a), the transcriptional regulation of which is dealt with below (see p. 116). It is usual to refer to the different poly-peptidecoding sequences of an operon as different genes, with the implication that mutations in one are without effect on the expression of another.

It is customary to say that operons do not occur in eukaryotes, but that statement needs qualification. The entire mitochondrial genome, at least in rats and humans where the research has been done, turns out to be a single transcriptional unit, even though it encodes numerous different products including ribosomal RNA (rRNA) and tRNA molecules as well as proteins. The difference from a bacterial operon is that the single long primary mitochondrial transcript is fragmented before translation. It has punctuation marks in the form of precursor sequences for mitochondrial tRNA molecules (see Fig. 5.18); the excision of these tRNA precursors also separates the mRNAs for the mitochondrially encoded proteins. Though the mammalian mitochondrial genome is a single unit of transcription, the mRNAs into which it is processed are translated quite independently of one another.

In bacterial operons, on the other hand, the transcript encoding different proteins is not pro-cessed into separate messengers before translation. Consequently, within an operon, the translation of one open reading frame ('gene') is not necessarily independent of that of another. Chain-termination mutations in one gene often have markedly de-pressing effects on the levels of translation of genes further downstream. The earlier the chain termination in the translation of the upstream gene, the stronger the effect on downstream gene expression (Fig. 4.4b). This polar effect has been shown to be due to destabilization of the operon messenger RNA, which is normally protected from ribonuclease attack by a more or less continuous train of ribosomes; the termination codon of one gene is followed closely by the initiation codon of the next. When ribosomes are shed prematurely as the result of a chain-termination mutation, the RNA will be unprotected to an extent dependent on the distance to the next initiation codon.

The more upstream the mutation, the longer the tract of exposed mRNA, and the lower the level of surviving downstream messenger.

Operons are therefore not just units of tran-scription − they are to some extent integral units of translation as well. However, even polar *lacZ* mutants generally have enough *lacY* activity to complement *lacY* mutants. Thus, by the com-plementation criterion, the different polypeptide-encoding sequences of the *lac* operon (and other operons by the same argument) have always been accorded the status of separate genes. In eukaryotes, except for mitochondrial genomes, it is more straightforward to use the term gene to refer to the unit of transcription.

The gene as a unit of transcription

Genomic DNA compared to cDNA − the detection of introns

As we saw in Chapter 3, gene transcripts can be cloned as cDNA, and the genomic sequences from which they are transcribed can be recovered from genomic 'libraries' using cDNA probes. One might expect the genomic and cDNA sequences to be identical.

The most graphic way of showing whether or not this is the case is by electron microscopy. The cloned cDNA and the corresponding cloned genomic DNA are made single stranded, mixed and allowed to reanneal. The reconstituted DNA molecules, some of which will be hybrids, are generally mixed with cytochrome c, a readily available and strongly basic protein that sticks to the acidic DNA and covers it with a thick protein coat to make it easier to visualize. The results in many cases are contrary to simple expectation. The two sequences do indeed anneal, but the genomic sequence may be longer − sometimes very much longer − than the corresponding cDNA. The extra length of the genomic sequence is accom-modated in the hybrid molecules through looping-out as single strand at particular sites. Figure 4.5 shows an example.

A priori, there would be two ways of explaining this unexpected relationship between the gene and its cDNA. Either the mRNA, of which the cDNA was a copy, is transcribed from several different

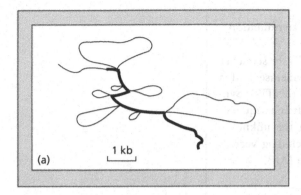

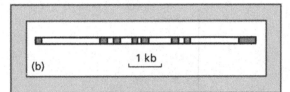

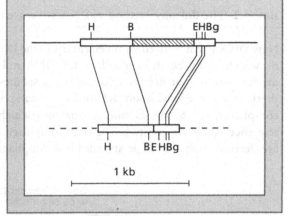

Fig. 4.6 Comparison of restriction-site maps of genomic and cDNA clones of the rabbit β-globin gene. The cross-hatched large intron is clearly missing from the cDNA sequence. A second, much smaller, intron was missed in this analysis. H, B, E and Bg stand for *Hin*dIII, *Bam*H1, *Eco*RI and *Bgl*II restriction sites. Based on Jeffreys & Flavell (1977).

Fig. 4.5 (a) An example of detection of introns by heteroduplex formation. A cloned genomic fragment from chicken, containing the ovalbumin gene, was hybridized to the corresponding cDNA by denaturing the mixture to single strands and reannealing. After coating with cytochrome c, duplex structure showed in electron micrographs as relatively thick rigid filaments. The thinner, more flexible loops are single-stranded introns not hybridizing to the cDNA. (b) The exon–intron map of the gene: exons stippled and introns open bars. Redrawn from Dugaiczyk *et al.* (1979).

gene segments and then spliced together (always in an order corresponding to the order of the corresponding segments in the genome), or the mRNA is shortened from the primary transcript by the excision of certain segments and the joining of the flanking cut ends. For reasons that will become clear, the second alternative, often referred to as 'splicing out', is the correct one. The RNA segments that are spliced out of the primary transcripts are called *introns*, and the segments that are retained in the mRNA are called *exons*.

Comparison of genomic and cDNA restriction-site maps

Another way of comparing the structure of a gene with a cDNA copy of its transcript is through their respective restriction-site maps. The first case in

which such a comparison was made was that of the rabbit gene for the β-chain of globin (Fig. 4.6). The results demonstrated an intron of approximately 500 bases – a second, smaller, intron went undetected in this analysis. The same comparison defines, within the level of resolution of the map, the end-points of the transcription unit in the genomic sequence, but without showing which is the initiation end and which the termination end.

Introns defined by sequence analysis

Obtaining a restriction-site map of a gene is the first step towards defining its complete nucleotide sequence. Overlapping fragments of the map are cloned into a plasmid or bacteriophage vector which is usually especially adapted for sequence determination through having a sequencing primer site just upstream (in terms of the direction of primed replication) of the fragment insertion site. The principle of the universally used dideoxy sequencing procedure, and the use of a sequencing vector, are explained in Box 4.2.

Provided one knows the amino acid sequence of the gene translation product, comparison of amino acid sequence and DNA nucleotide sequence will

Box 4.2 The dideoxy method of DNA sequence determination

New DNA is synthesized complementary to the DNA the sequence of which is to be determined, using DNA polymerase and a mixture of the four deoxynucleoside triphosphates (dNTPs). Synthesis starts at a fixed point defined by a primer, often a sequence complementary to the cloning vector into which the unknown sequence has been inserted. The commonly used cloning vectors are derived from the single-stranded bacteriophage M13.

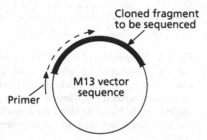

Four DNA polymerase reactions are run in parallel; each reaction contains the dideoxy analogue (ddNTP) of one of the four dNTPs as well as the normal dNTPs, one of which is labelled with ^{32}P to make the DNA products radioactive. When a dideoxynucleotide is incorporated, further chain elongation is impossible.

The normal 2′-deoxy residue carries a 3′-hydroxyl group for addition of the next residue

dNTP

The 2′,3′-dideoxy residue provides no 3′-hydroxyl, so chain elongation stops

continued

Box 4.2 *Continued*

For example, in a reaction containing a suitable ratio of ddCTP to dCTP, chain elongation terminates, with a certain small probability, wherever the polymerase incorporates a C opposite a G on the template strand.

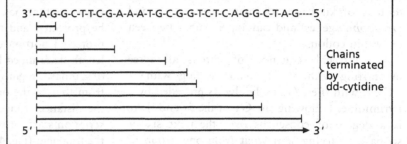

Chains terminated by dd-cytidine

The mixture of terminated single-stranded chains are melted off the template and separated by size on a long electophoresis gel, which is then dried and exposed to X-ray film. This reveals a ladder of radioactive DNA single-stranded fragments, up to 500 on one gel, each rung in the ladder corresponding to one residue in the template DNA. Each rung is present in just one of the four tracks – the one with the ddNTP that terminates extension at the corresponding position in the DNA sequence.

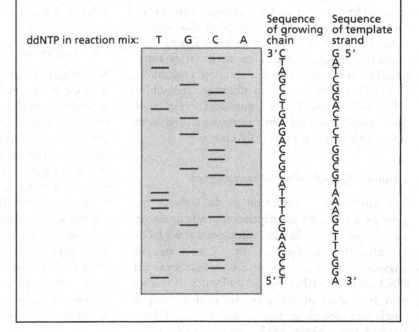

show at once whether the coding sequence in the DNA is continuous or interrupted by introns. Genes without introns have unbroken open reading frames from an initiation (ATG) codon to a termination (TAA, TAG or TGA) codon, with codons in between corresponding precisely to the known amino acid sequence. Introns are of very variable length (from 50 to 100 bases in fungi to tens of kilobases for many mammalian and *Drosophila* genes) and can appear either between or within codons.

The internal sequences of introns all agree in starting with 5'-GT··· and ending with ··· AG-3', with the AG nearly always preceded by a pyrimidine. Following the GT at the 5' end there is a degree of consensus over the next six or so bases, varying somewhat from one group of organisms to another. At the 3' end there is very little consensus upstream of the AG except for a tendency to short runs of pyrimidine residues. However, in the interior of the intron, usually within a few tens of bases of the 3' end, there is a sequence that plays an important part in the splicing-out process (providing the branch-point or 'lariat' junction sketched in Box 4.2). This sequence, in *S. cerevisiae*, is always 5'-TACTAACA-3'. In most other eukaryotes one tends to find shorter and less constant versions of this sequence (e.g., 5'-CT$^{A}_{G}$AC-3' in filamentous fungi and to some extent in animals). Given these common, though limited, earmarks, it is often possible to identify with fair certainty reasonably short introns interrupting an otherwise satisfactory ORF, and so deduce the amino acid sequence of the product encoded after intron excision, even where this was not previously known.

Defining the end-points of transcription

The more precise definition of the ends of the gene as a unit of transcription depends again on nucleic acid hybridization. Single-stranded DNA is vulnerable to digestion by S1 nuclease, an enzyme that does not attack double-stranded DNA or DNA–RNA hybrid molecules. If restriction fragments of the gene are cloned using a single-stranded phage vector, such as one of those derived from phage M13, the fragment can be amplified in radioactive single-stranded form. Either strand of the original duplex can be amplified depending on the orientation of the insertion into the vector. The appropriate DNA single strand can then be annealed with a large excess of mRNA including the processed transcript of the gene under investigation, and the annealed mixture digested with S1 nuclease. The DNA sequence that has hybridized with mRNA will be protected and the rest will be digested. The protected portion can then be recovered and its length determined by comparison with a 'ladder' of standard fragments. That length will be reduced from that of the original restriction fragment by the distance from the fragment end to the transcription start- or stop-point, if either falls within the fragment (Fig. 4.7). In this way the ends of the transcribed unit can be placed within the genomic restriction map.

Processing of the primary transcript

RNA polymerases – promoters and processing of transcripts

In eukaryotes, there are three kinds of RNA polymerase, all multisubunit complexes, serving three different functions (Fig. 4.8).

RNA polymerase I (PolI)

PolI transcribes the genes for rRNA which, as we shall see in the next chapter, are always present in multiple tandemly arrayed copies. A single primary transcript is cleaved into the large (28S) and small (18S) rRNAs for the large and small ribosomal subunits and a smaller (5.8S) fragment that forms part of the large subunit. Here, then, we have a single gene (in the sense of a unit of trancription) providing three RNA products.

Initiation of transcription by PolI depends on a special promoter sequence situated a little way upstream of the transcriptional startpoint, generally in the untranscribed spacer between successive rRNA genes. It is the sequence to which the polymerase and other accessory proteins bind to initiate transcription.

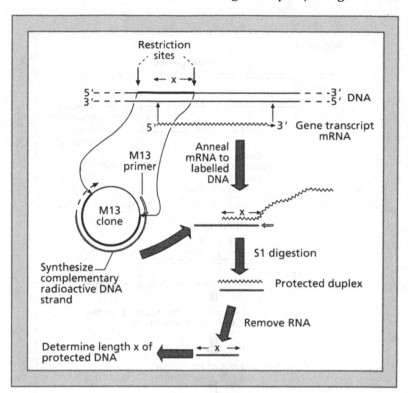

Fig. 4.7 S1 nuclease mapping of transcription end-points. In the procedure depicted, a restriction fragment overlapping the ends of the transcribed sequence is cloned in single-stranded bacteriophage M13 and hybridized to messenger RNA. Non-hybridized tails are digested away with the single-strand exonuclease S1 and the protected DNA–RNA heteroduplexes are isolated. The protecting RNA is then enzymically removed and the lengths of the remaining single-stranded DNA fragments determined by gel electrophoresis alongside a 'ladder' of sequences of known sizes. These lengths then represent the distances between the transcription end-points and the relevant restriction sites.

RNA polymerase III (PolIII)

PolIII is responsible for the synthesis of many different small RNA molecules, including tRNAs and also the components of a range of ribonucleoprotein particles including the 7SL RNA of the signal recognition particle (see pp. 137–138). The extent to which these small RNAs are processed after transcription varies from one case to another. They all have extensive secondary structure in the form of stems and loops stabilized by base-pairing, and in some cases a stem—loop structure formed by the primary transcript is not present in the mature RNA molecule.

PolIII promoters have the special and significant (see p. 138) property of being located downstream of the transcriptional startpoint. The promoter is also unusual in being bipartite, consisting of two approximately 20 bp sequences with a space of about 20 bp in between. It is thought that two accessory trancription factors bind to the promoter

initially and form a complex with the large, multisubunit PolIII molecule, which contacts the transcription startpoint.

RNA polymerase II (PolII)

PolII transcribes protein-encoding genes. PolII promoters are located upstream of the transcriptional startpoint. Their various and complex structures and functions are dealt with in Chapter 7.

PolII transcripts are processed in three different ways to make mature mRNAs – 'capping', 'tailing' and intron excision. The protective 5' cap group is a methylated guanosine triphosphate derivative, attached the 'wrong way round', 5'-to-5'. The 3' tail consists of a polyadenylate (poly-A) sequence of up to a few hundred adenylate residues. The poly-A addition is prompted by a special polyadenylation signal at the 3'-end of the transcript; in animals it is 5'-AAUAAA-3' but is

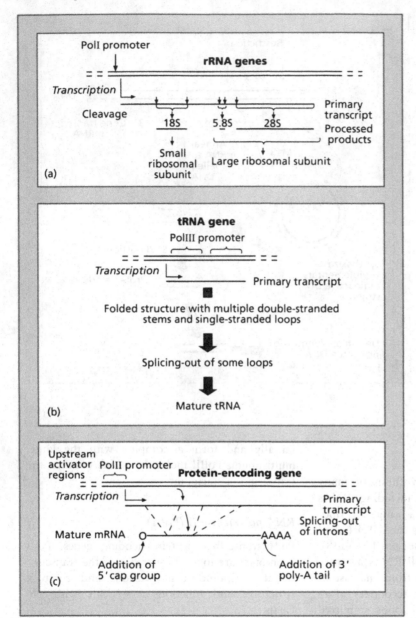

Fig. 4.8 Processing of eukaryotic transcripts synthesized by different RNA polymerases. (a) Ribosomal RNA (rRNA), transcribed from an upstream promoter by PolI, cleaved to form different functional products. (b) Transfer RNA (tRNA), transcribed from an internal promoter by PolIII, formed by trimming of some loops formed by the primary transcript, with some base modification. (c) Messenger RNA (mRNA), transcribed from an upstream promoter by PolII, followed by capping, tailing and splicing-out of introns.

somewhat variable in other organisms. The transcript is cleaved a short distance downstream of this signal sequence and the chain of As is added to the cut 3′ end. The poly-A tail gives eukaryotic mRNAs the useful property of being trappable on columns containing the complementary oligo-dT bound to cellulose or some other inert matrix (cf. p. 82).

The splicing-out of introns

This most intriguing mode of processing of PolII transcripts, is dependent on three critical nucleotide sequences within the introns themselves — the extreme 5′ and 3′ ends and the branch-point or 'lariat' sequence towards the 3′ end (Box 4.3). It also depends on the presence of all the com-

Box 4.3 Main steps in the splicing-out of introns

The splicing agents are a set of at least six ribonucleoprotein particles, each containing a small RNA molecule together with protein components, and called U1, U2, etc. Together they form an aggregate called the spliceosome. Only the functions of U1 and U2 are shown in the simplified diagrams below.

1 The essential sequences of a *Saccharomyces* intron, with (a) a consensus at the 5′ end, with only the initial GT absolutely invariant, (b) an invariant UACUAACA branch-point sequence, usually within 10 to 30 bases of the 3′ end, (c) an invariant AG, almost always preceded by a pyrimidine, at the 3′ end. The invariant 5′ and 3′ bases are present in all introns of eukaryotic nuclei, but the branch sequence shows a much looser consensus in organisms other than yeast; for example, other fungi and animals usually just conserve CUPuAC, where Pu is either A or G.

'Engineered' mutations (see Box 4.4) of a yeast intron sequence have been used to show that the specific bases marked ● are essential for splicing-out of the intron and that changes in those marked ○ greatly reduce splicing efficiency.

Sequences close to the 5′ ends of U1 and U2 RNA appear capable of hydrogen bonding to the intron 5′-end and branch sequences, respectively.

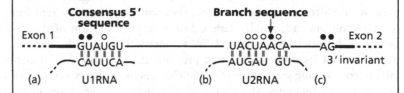

2 (*see figure overleaf*) (a) The 5′-end and branch sequences are brought into proximity by their binding to U1 and U2, which form a complex.

(b) The 2′ ribose hydroxyl group of the branch-point adenosine residue, arrowed above, forms a bond with the 5′ terminal guanosine residue to form a 2′-to-5′ branch, thus detaching the 5′ end of the intron from exon 1 and leaving the intron attached to exon 2 as a loop ('lariat').

(c) The free 3′-hydroxyl of exon 1 attacks the 5′ carbon at the terminus of exon 2, leading to ligation of exons 1 and 2 and the exclusion of the intron as a free lariat which is subsequently degraded.

The cleavage and ligation steps are catalysed by other components of the spliceosome. See Ruby & Abelson (1991).

continued on p. 110

Box 4.3 *Continued*

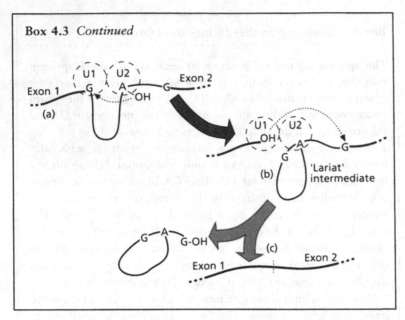

ponents of a complex particle, the *spliceosome*, which is made up of a number of ribonucleoprotein subunits each containing a small RNA molecule. Loss of activity of specific genes is never due to mutations affecting the spliceosome components; these would affect so many genes that they would certainly be unconditionally lethal. But a single gene can be effectively silenced by a mutation within one of its own introns that prevents that intron from getting spliced out. Certain changes in the conserved sequence at the 5′ end of introns result in accumulation of the unspliced primary transcript, while some at the 3′ end lead to accumulation of the 'lariat' intermediate, thus confirming the main steps of the splicing pathway as set out in Box 4.3. The evidence comes mainly from the study of splicing-defective mutants, especially from a series of 'engineered' changes in *Saccharomyces* introns, one of which is explained in Box 4.4. We shall see several other examples of the use of *in vitro* DNA manipulation as a means of exploring gene function.

The first idea about the significance of introns was that they were, so to speak, the scars left behind by gene fusions that had occurred during evolution. The exons of each subdivided gene had, it was suggested, been originally assembled from distinct genes encoding different but complemen-

tary protein domains, and intron-excision from their joint transcript had evolved as a means of fusing their respective open reading frames. Whatever the merits of this idea about intron origins, it is now clear that at least some introns have current functions not just confined to facilitation of their own removal. They have implications for the timing of transcription (see p. 205), and they sometimes contain enhancer sequences necessary for normal gene expression (see p. 202). Even more interestingly, they may permit the same gene to perform more than one function.

Different products from the same primary transcript

Genes with introns can have a degree of flexibility in their expression that is denied to intron-less genes. A good example is provided by a gene that has been identified in chicken, mouse, rat and human genomes and encodes a set of proteins involved in the mutual adhesion of neural cells. In all of these vertebrate species the gene has a complex pattern of exons and introns which are subject to different modes of splicing to generate a family of different, though overlapping, mRNAs. The situation in chicken, as evaluated by comparisons of different cDNA clones with the

Box 4.4 A procedure for making controlled mutations *in vitro*.
After Norris *et al.* (1983)

1 The gene sequence to be mutated is isolated as a restriction fragment and cloned into the polylinker site of a vector derived from bacteriophage M13. The clone is propagated in *E. coli* to make working quantities of single-stranded closed-circular DNA, which is the normal form of the DNA in infective M13 particles.

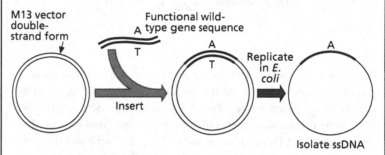

2 An oligodeoxynucleotide (shown here as an 8-mer, but it will usually be longer) is made so as to be complementary to the single-stranded insert but with one centrally located mis-match corresponding to the mutation it is desired to introduce. The oligonucleotide, and also another short sequence complementary to the M13 sequence, are annealed to the cloned DNA.

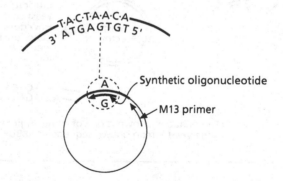

3 (*see figure overleaf*) The annealed oligonucleotides are used a primers for DNA synthesis catalysed by DNA polymerase, so that the whole insert and its adjacent M13 sequences are made double-stranded. The double-stranded hybrid (heteroduplex) insert is excised by appropriate restriction enzymes, and introduced into an *E. coli* plasmid vector carrying a selectable antibiotic-resistance marker. The plasmid, with its insert, is introduced into *E. coli* and antibiotic-resistant colonies are isolated. As the plasmid replicates, a half of the daughter plasmids will become fixed for the mutation.

continued on p. 112

Box 4.4 *Continued*

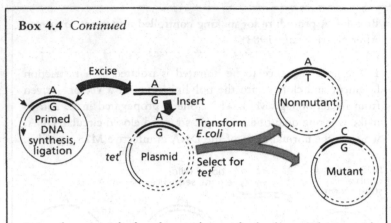

4 A colony is picked and spread on a fresh plate. Colonies with plasmids carrying the mutant gene sequence are distinguished by colony hybridization (see Fig. 3.11) with a probe specific for the mutant sequence. Such colonies are propagated and used to prepare plasmid DNA. The mutant insert is cut out using the same restriction enzymes as before and inserted into a yeast shuttle vector for functional testing in yeast.

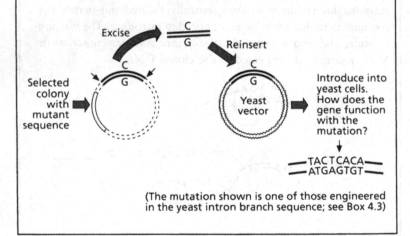

(The mutation shown is one of those engineered in the yeast intron branch sequence; see Box 4.3)

genomic sequence, is shown in Fig. 4.9. A striking feature of the gene is the large number of exons (at least 23), some of them extremely short, separated by much longer introns. The exons add up to about 3.2 kb, spread over approximately 46 kb of genomic sequence.

In chicken brain, three neural cell adhesion molecules (N-CAMs) of different sizes have been detected, corresponding to three kinds of mRNA, cloned as cDNA. As shown in Fig. 4.9, the largest mRNA molecule includes all the exons except exon 15. The next largest one excludes exon 18 as well as exon 15, while the smallest one includes exon 15 but terminates translation at the end of that exon. In cDNA clones from heart and skeletal muscle a further source of variation was discovered – the inclusion in some mRNAs but not in others of four previously unrecognized exons between exons 12 and 13: 12A, 12B, 12C and 12D, only 14, 33, 42 and 3 bases long.

Thus a single gene, in the sense of one unit of transcription, can encode a whole family of

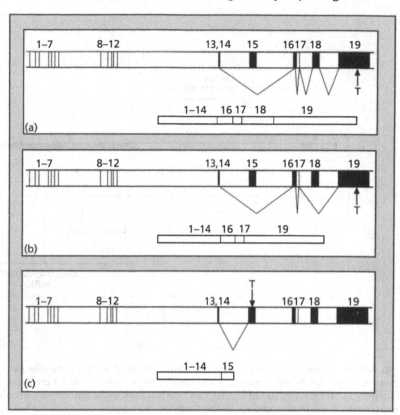

Fig. 4.9 Alternative modes of splicing of the transcript of the chicken gene encoding neural cell adhesion protein (N-CAP). The exons, numbered 1–19, are shown as black bars and the open boxes between them are the introns. T indicates the positions of termination codons. Three different modes of splicing, (a), (b) and (c), generate three messenger RNAs of different lengths and exon compositions, shown below the splicing diagram in each case. The (a) and (b) products encode polypeptides with a membrane-spanning sequence, encoded by exon 16, connecting sequences that are outside and inside the cell, respectively. The (a) product has a more extensive intracellular domain than the (b) product, and the (c) product is thought not to span the membrane but just to attach to the cell surface. After Cunningham *et al.* (1987).

proteins depending on the mode of intron splicing. The different protein products, with the exception of the smallest which is confined to the cell surface, span the outer membrane of neural cells, with exons 1–15 corresponding to the extracellular domain, exon 16 the membrane-spanning sequence, and exons 17–19 the domain within the cell. It is likely that the variations, which to some extent seem to be tissue-specific, play a vital role in the specification of cell-to-cell connections during neural development.

The *D. melanogaster* transposable P-element (cf. p. 133) provides a clear example of tissue-specific splicing. The element has a single primary transcript including a single ORF interrupted by three introns, encoding a protein product necessary for P-transposition. The third intron is spliced out only in the germ line, and consequently P element movement is normally detected only in germ cells, not in somatic tissues. More importantly for the fly, the whole pathway of *Drosophila*

sexual development is controlled by differential intron splicing (see p. 193).

Genes nested within other genes' introns

Of all the unexpected properties of gene introns, perhaps the most bizarre is the finding, in a few cases, of other genes 'nested' within them. The *Drosophila* gene *Gart* is one of the best examples. This gene is exceptionally interesting in illustrating three kinds of complexity simultaneously (Fig. 4.10). Firstly, it is one of the *Drosophila* examples of a gene responsible for several different enzymic functions (cf. p. 99). Its long polypeptide product catalyses three different steps in purine synthesis. The enzyme activities, abbreviated to GARS, AIRS and GART, are, in *E. coli*, the responsibility of three separate enzymes encoded by distinct genes. Secondly, it illustrates another way of getting more than one processed transcript from the same gene. The fourth intron contains a poly-

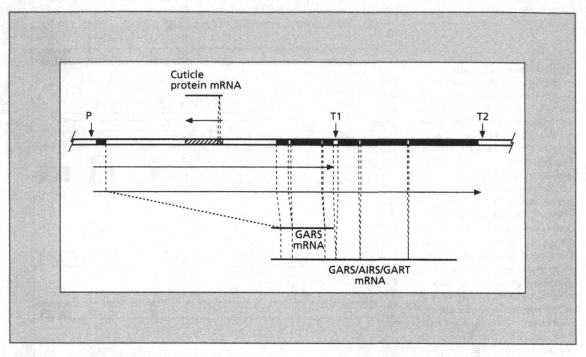

Fig. 4.10 An example of a gene 'nested' within the intron of another gene. The *Gart* gene of *Drosophila*, transcribed and translated from left to right, encodes a trifunctional enzyme responsible for three enzyme activities in the pathway of purine biosynthesis, abbreviated to GARS, AIRS and GART. *Gart* exons and introns are shown as stippled and empty bars, respectively. Because of a partially effective transcription termination sequence (poly-adenylation site, T1) in the fourth intron, two messenger RNAs (mRNAs) are produced, the smaller one encoding only GARS activity. The long first intron contains a gene for cuticle protein, shown as a hatched bar. It itself contains a small intron, and is transcribed and translated from right to left. Thus *Gart* uses one DNA strand as template and the cuticle protein gene uses the other. P = promoter; T1, T2 = transcription terminators. After Henikoff *et al.* (1986).

adenylation signal that terminates a proportion of transcripts to give a truncated mRNA encoding only the GARS function, but allows some transcription to proceed into the AIRS and GART regions.

However, what mainly concerns us in the present context is the third complication – the presence, in the very long first intron, of another gene, oriented in a direction opposite to that of *Gart*. This 'nested' gene encodes the polypeptide sequence of a cuticle protein, an essential component of the protective coat for the pupa, within which the larval tissues are replaced by the structures of the adult fly. This cuticle protein gene itself contains a short intron, ruling out the possibility that it is a processed pseudogene arising from the integration of a reverse-transcript as described in the next chapter (see p. 139). The nested gene was shown to be transcribed at the time of pupation.

Within *Gart*, different protein products are encoded by opposite strands of the double helix, and thus by different transcripts. Presumably, since both strands of a DNA duplex could hardly be transcribed at the same time, the two *Gart* transcripts are produced in different tissues or at different stages of development. This kind of dual usage of DNA would hardly be possible if the exons were overlapping, since in that case the amino acid sequence of one protein would be arbitrarily constrained by that of the other. The problem is avoided by the positioning of the cuticle protein exons entirely within a *Gart* intron.

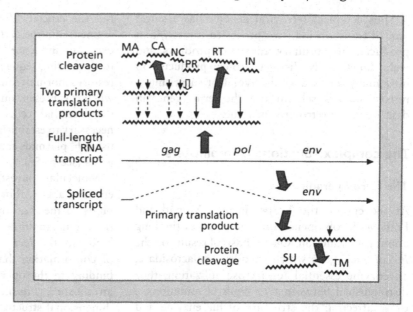

Fig. 4.11 Multiple products of the long primary transcript of a typical retrovirus proviral (chromosomally integrated) sequence. Incomplete splicing generates both full-length and *env*-only messenger RNAs. Because of a partly effective barrier to translation (⇕ either a frameshift or a termination codon) between the *gag* and *pol* sequences, the full-length messenger is translated to give both *gag/pol* and truncated *gag*-only products. The *env*, *gag* and *gag/pol* products are all polyproteins, cleaved (↓) by protease (PR) into separate proteins with different functions in the virus life cycle (see text). Based on the review by Varmus & Brown (1989).

Retroviral genome expression – units of all kinds!

Retroviruses, which are almost ubiquitous in animals and also present in some plants, have RNA rather than DNA in their infective particles. After infection into the host cell, the RNA is reverse-transcribed into DNA by reverse transcriptase, an enzyme present in the virus particles. The reverse transcript is made double-stranded and integrated into the host genome as a *provirus*. We return to retroviruses in the context of the whole genome in the next chapter. Here we are concerned just with the mode of expression of the provirus (Fig. 4.11).

Virtually the entire length of the provirus sequence is transcribed as a single unit, conventionally described as consisting of three functionally distinct regions named for some of the main proteins that they encode – *gag* (group-specific antigen), *pol* (polymerase, i.e., reverse transcriptase) and *env* (virus envelope protein). A proportion of this long transcript is processed by splicing-out of the *gag/pol* regions, leaving only the downstream *env* domain. The unprocessed fraction contains all three regions. Presumably these different messengers are required in different proportions for the fabrication of infective virus particles.

The separate *env* messenger is translated into a polypeptide chain that is subsequently cleaved to two products, SU (surface protein) and TM (transmembrane protein, for penetrating the host cell membrane). The full-length messenger is translated into a gal–pol *polyprotein*, so-called because it is cut into a number of functionally distinct polypeptides. The boundary between the *gag* and *pol* parts of the messenger is marked, in different retroviruses, by either a chain termination codon or a reading-frame shift. Through special mechanisms that need not concern us in detail, these obstacles are partly overcome, the result being a mixture of full-length gag–pol polyprotein and a larger amount of the truncated gag-only product. Again, the ratios of the alternative products reflects the different requirements for the various protein products; the gag products are structural proteins and the pol products enzymes, needed only as catalysts. The gag polyprotein is cleaved to make four inner components of the virus particle as shown in Fig. 4.11, and pol yields reverse transcriptase (RT) and a protease (PR), the latter presumably involved in polyprotein processing.

Thus a typical retroviral genome has a single unit of transcription, two processed transcription products, three primary translation products, and eight functionally distinct protein products. So how many genes does it have? Not surprisingly, retrovirologists seldom use the term 'gene' in describing the retrovirus genome.

The complex functions of promoters

The *E. coli* paradigm

At the end of the 1950s, Jacques Monod and Francois Jacob opened up a new way of thinking about gene function. On the basis of results on the *E. coli lacZ* gene which encodes β-galactosidase, an enzyme essential for lactose utilization, they distinguished between two kinds of mutations – those affecting the structure of the enzyme and those affecting the quantity of enzyme produced under various conditions of culture. In the latter class they detected two different kinds of mutants which, by recombination analysis, they mapped in two short adjacent segments, the *promoter* and the *operator*, immediately upstream of the structure-determining part of the gene (see Fig. 4.4).

Mutations in the promoter decrease, sometimes drastically, the maximum amount of β-galactosidase that can be produced. This is due to a general reduction in the efficiency of transcription. The promoter was thought to be the transcription start-point – presumably the binding site for RNA polymerase.

The operator mutations disrupted the normal nutritional control of transcription. Whereas, in wild-type *E. coli*, β-galactosidase mRNA is induced by lactose in the growth medium and repressed by glucose, in certain operator mutants it is formed constitutively, that is at a high level regardless of the sugar providing the carbon source for growth.

The operator, like the promoter, acts on the structure-determining sequence only in *cis*. There is, however, another component of the system of transcriptional control that acts in *trans*. Mutations in a linked, but functionally quite independent gene, called *lacI*, give the same constitutive phenotype as the operator mutations.

But the crucial difference between the operator segment and the *lacI* gene is that the former only exerts its influence on *lacZ* when actually joined to the coding sequence (i.e., in *cis*) whereas *lacI*+ restores normal control to a *lacI*− mutant even when it is present on a separate DNA fragment in the bacterial cell. The *cis–trans* comparisons necessary to establish these conclusions were made using F′ plasmids carrying second copies of the *lac* locus (see p. 52).

Molecular investigation shows that *lacI*+ encodes a DNA-binding protein that specifically binds to the *lacZ* operator, inhibiting transcription. A metabolic derivative of lactose, allolactose, binds to the *lacI*+ protein, inducing a change of conformation that prevents the protein from binding to the operator. This is an example of an *allosteric transition* – a change in the three-dimensional structure of a protein in response to the binding of a smaller molecule.

Examples of this kind of system, with the *trans*-acting protein product of one gene binding to the *cis*-acting segment of another, are numerous both in prokaryotes and in eukaryotes. However, in many of the prokaryotic systems and a majority of the eukaryotic systems that have been analysed, the *trans*-acting protein acts as an *activator* of transcription rather than a repressor (or sometimes as either an activator or a repressor depending on the circumstances). Frequently, as we see below, more than one transcriptional activator is necessary for fully regulated gene transcription.

With the extension of studies on gene regulation to other organisms and other systems, Jacob and Monod's original distinction between promoter and operator has not been found useful. The word promoter is now used to refer to the whole upstream segment concerned with assembly of a transcription initiation complex, including an RNA polymerase (itself a multisubunit protein) and various other protein factors that assist (or restrict) its function. Some of the components of the polymerase II initiation complex are common to many, if not all, genes. One of these is transcription factor IID (TFIID), itself a complex of proteins one of which binds to a sequence immediately upstream of the translation start-point that usually approximates to TATAA and is commonly known as the 'TATA box'. TFIID is a general and unspecific factor; specificity is often provided by successive binding of other proteins, some of which, as we shall see, are specific for particular genes or sets of genes.

Analysis of eukaryotic promoter regions by mutation

The geneticist's approach to the analysis of a functional mechanism – to change its components one at a time by mutation and to see what happens – has been successfully applied to promoter function. Useful promoter mutations have sometimes been obtained by *in vivo* mutagenesis but, in recent work, the mutations have mostly been manufactured *in vitro* and introduced artificially into living cells or, preferably, into whole organisms. In this section we shall consider three examples from different organisms to illustrate different experimental approaches to the problem of promoter structure and function.

The acetamidase (amdS) gene in Aspergillus nidulans

The *amdS* gene of *A. nidulans* encodes the enzyme acetamidase, which hydrolyses acetamide to acetic acid and ammonia. It is necessary for the utilization by the fungus of acetamide either as nitrogen or carbon source. Its transcription is regulated by at least four different *trans*-acting genes, *areA*, *amdR*, *facB* and *amdA*, which are unlinked to *amdS* or to each other (Fig. 4.12). Specific sites have been identified within the *amdS* promoter region that critically affect the response to each of the last three.

areA acts positively to activate trancription of *amdS*, but only on growth media low in ammonium nitrogen. It seems that transcriptional activation by the *areA* protein product is negated by its binding to some ammonia derivative, probably glutamine. Regulation by *areA* extends to all genes encoding enzymes with the function of supplying ammonium nitrogen. Presumably all these genes have a specific binding site for the

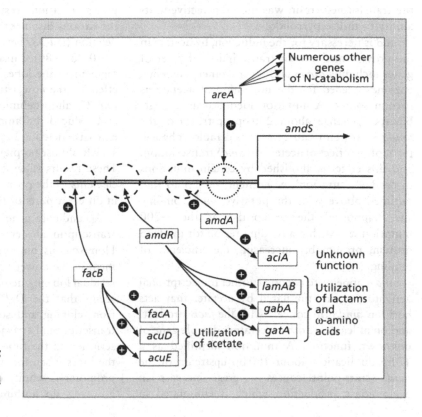

Fig. 4.12 Multiple positive regulation of the *amdS* (acetamidase) gene of *Aspergillus nidulans*. The *trans*-activating proteins regulate overlapping sets of genes. Regions of the promoter region, all within 200 bases of the transcription start, that bind the *facB*, *amdR* and *amdA* products that activate transcription (⊕) are circled; the *areA* binding site has been only tentatively defined. The activators are not specific for *amdS* but also activate non-overlapping sets of other genes (see text). Based on Hynes & Davis (1986).

areA protein in their respective promoter regions.

amdR supplies a transcriptional activator that is effective in the presence of ω-amino acids, such as β-alanine and γ-aminobutyric acid. Genes that encode enzymes for the metabolism of these compounds, and also of some lactams such as pyrrolidone, are under *amdR* control – and so, for functional reasons that are not immediately obvious, is *amdS*. A mutation that has been identified as a deletion of 43 bp about 160 bp upstream of the *amdS* transcription start-point eliminates the stimulation of *amdS* transcription by ω-amino acids in the growth medium. The hypothesis that this mutation eliminates a binding site for the *amdR* protein product was supported by an ingenious experiment in which multiple copies of a DNA fragment including the suspected sequence were integrated into the genome through a transformation experiment. The prediction was that these additional copies of the binding sequence would 'titrate' most of the *amdR* product, leaving insufficient to support transcriptional activation of the genes involved in ω-amino acid and lactam utilization. The prediction was fulfilled, in that the transformed strain was indeed defective in its ability to utilize pyrrolidone.

facB is necessary for the induction, by acetate in the growth medium, of transcription of a set of genes including *amdS* and several others encoding enzymes needed for the utilization of acetate as carbon source. A mutation identified as a single base-pair change about 200 bp upstream of the *amdS* start-point causes a considerable enhancement of the effect of acetate on *amdS* transcription, and this effect is abolished by certain mutations in *facB*. A 'titration' experiment, similar to that outlined above with the putative *amdR* binding site, supported the proposal that the −200 mutation was within a binding region for the *facB* protein, presumably increasing the efficiency of binding.

amdA appears to encode another transcriptional activator, again stimulated by acetate, that acts both on *amdS* (in concert with the *facB* product) and on at least one other acetate-inducible gene of unknown function. A mutation identified as a 17 bp duplication about 100 bp upstream of the *amdS* start-point causes a great increase in acetamidase production which is eliminated in

amdA mutants. It is inferred that the mutational insertion defines a binding site for the *amdA* product.

The *Aspergillus amdS* gene is one of the better-analysed examples of a complex system of transcriptional controls concentrated within just a few hundred bases of upstream sequence, but it is certainly not exceptional. This kind of complexity of promoter function is turning out to be very much the rule for a wide range of eukaryote genes.

The pallida (Pal) gene of Antirrhinum

The *Pal* gene of the garden snapdragon, *Antirrhinum majus*, encodes the enzyme responsible for the final step in the synthesis of the magenta anthocyanin flower pigment, cyanidin. Homozygous null *pal* mutants have ivory flowers. A strain characterized by spots and stripes of magenta on an ivory background was shown to be carry a transposable element (*Tam3*) immediately upstream of the transcription startpoint. The fully coloured sectors were due to more-or-less precise excision of *Tam3* during flower development; imprecise excisions resulted in deletions extending for various distances into the promoter region. A deletion (*pal-33*) between 70 and 80 bp upstream (−70 to −80) caused a marked reduction of pigment in the lobes of the flower but had little effect on the flower tube. A −70 to −90 deletion (*pal-32*) almost eliminated pigment from the lobes and reduced the amount in the tube. The more extensive deletion between −70 and −170 in *pal-15* left almost no pigment either in lobes or tube. These results suggest that there may be separate elements in the promoter governing pigmentation in different parts of the flower.

A candidate gene for encoding a tube-specific transcriptional activator is called *delila* (*del*). Homozygous mutant *del* eliminates purple pigment in the flower tube, and strongly reduces the residual lobe pigmentation in *pal-32* and *pal-33*. It seems that the *Del*[+] product (which is known from cloning and sequencing to resemble other transcriptional activators) is essential for *Pal* transcription in the tube and helps *Pal* expression in the lobes when the main lobe-specific part of the promoter is wholly or partly deleted. *Del*[+] can, however, do nothing to help the *pal-15* allele,

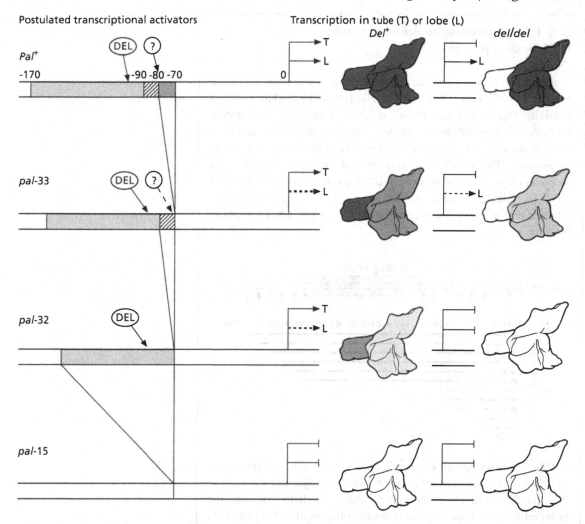

Fig. 4.13 Regulation of purple flower colour of *Antirrhinum majus* by different segments of the promoter of the *Pal* gene. The *Del*⁺ gene is required for transcription of *Pal*⁺ in the flower tube. The *Pal* promoter region within 170 bases of the transcription start (0) was deleted to different extents by imprecise excision of the *Tam3* element from its site of insertion in the *pallida-recurrens* allele (cf. Fig. 3.16). Deletion of the 10-bp region between −80 and −70 bp leaves a reduced level of pigmentation in the flower lobes, reduced still more in the absence of *Del*⁺. A 20-bp deletion between −90 and −70 bp reduces lobe pigmentation to a very low level, now completely dependent on *Del*⁺. The longer deletion extending all the way back to −170 bp eliminates all purple pigment, irrespective of presence or absence of *Del*⁺. The hypothesis is that the *Del*⁺ product (DEL), known to be a DNA-binding protein, binds upstream of −90 bp to activate *Pal* transcription in the tube and, to some extent, in the lobes as well; and another unidentified protein (?) is thought to bind at −90 to −70 bp or (less effectively) to −90 to −80 bp alone to activate transcription in the lobes. After Almeida *et al.* (1989).

presumably because *pal*-15 has lost the DEL-binding site (Fig. 4.13).

This second example, like the first, depends on mutants generated fortuitously *in vivo*. In our other example, however, the crucial information was obtained by controlled *in vitro* manipulation of promoter sequences.

The ovalbumin gene of chicken

Ovalbumin, the predominant egg-white protein, is produced exclusively in the egg-producing tissue of the oviduct, and the gene that encodes it is transcribed in response to a number of signals including the female hormone oestrogen. A se-

Box 4.5 'Footprinting' to define the binding site for a DNA-binding protein

· ·

A restriction fragment including the sequence of interest is labelled at one 5'-end with ^{32}P (* in the diagram below). The DNA is partially digested with DNaseI in the presence of the binding protein. DNaseI cleaves at random between any two successive nucleotide residues, except where the DNA is protected by binding to protein. The result of a very limited digestion, followed by removal of protein and melting of DNA to single strands, will be a series of radioactive single-stranded fragments extending from the labelled end to every position in the chain except within the DNA-binding region.

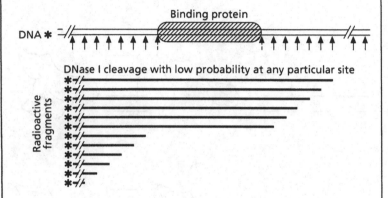

When run out on an electrophoresis gel, the labelled fragments will form a ladder with a region of missing rungs. The sizes of the missing digestion products, corresponding to distances from the labelled 5'-end of the original restriction fragment to the protected region, are determined by reference to a ladder of fragments of known sizes. This defines the part of the sequence that is covered by the binding protein.

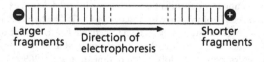

quence 40–50 bases upstream of the transcription startpoint and just 16 bases upstream of the TATA 'box' has been identified as the binding site for the oestrogen receptor, a protein that forms a stable complex with the hormone and then helps to activate transcription of a set of genes expressed in the oviduct.

The identity of the receptor-binding sequence in the DNA was established by a 'footprinting' assay (Box 4.5) and its function was explored by testing the ability of a series of DNA constructs with different base-pair replacements to promote transcription when introduced into cells. In this experiment, as is often the case in this kind of work, the gene whose expression was monitored was not the one that is naturally attached to the promoter region under test, but rather a 'reporter' gene whose product was quicker and easier to measure quantitatively. A favourite reporter is the bacterial gene encoding *chloramphenicol transacetylase* (*CAT*), an enzyme with the natural function of conferring resistance to the antibiotic chloramphenicol by acetylating and inactivating it.

In the series of constructions diagrammed in Fig. 4.14 the entire coding sequence of the *CAT* gene, separated from its own promoter, was fused in the proper orientation to the ovalbumin gene control region. The variously modified sequences were introduced (see Table 3.2) into cultured chicken fibroblast cells, and the amounts of CAT produced were measured after 48 hours. This is a system called *transient expression*; it does not rely

on integration and replication of the introduced DNA but just on the short-term transcription of those constructs that persist for a while in the cell nucleus.

A sequence extending only 58 bases upstream of the transcription start-point did indeed promote transcription and synthesis of CAT in the presence, but not in the absence, of oestradiol in the cell culture medium. Base-pair changes either in the TATA box or in the GGTCA hormone receptor-binding sequence (respectively around 30 and 46 bases upstream of the transcription start) brought about a three- to fivefold reduction in the amount of CAT synthesized (Fig. 4.14).

Other studies showed that this same short promoter sequence contained a binding site for the receptor of another hormone, progesterone, which also stimulates transcription. Here we have another example of a promoter region subject to control by multiple *trans*-acting proteins.

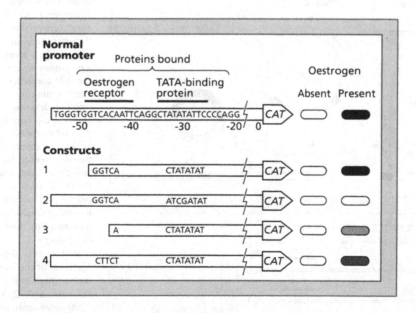

Fig. 4.14 Reconstruction of the promoter of the chicken ovalbumin gene to modify its response to hormone. Within 50 bases of the transcription start there are two protein-binding sequences. One, the TATA 'box' binds to TATA-binding protein, believed to be general transcription-activating factor TfIID. The second binds, in the presence of oestrogen, to oestrogen-receptor protein. The cloned promoter, fused to the *CAT* (chloramphenicol transacetylase) reporter gene was reconstructed as shown (1 and 3 truncated at the upstream end; 2 and 4 with sequence alterations) and introduced into chicken fibroblasts. Transient CAT activity was measured with and without administration of oestrogen to the cell cultures. CAT activities visualized as bands of radioactive reaction product on chromatograms, are indicated to the right. After Tora *et al.* (1988).

Enhancers – the expanding frontiers of the gene

The ADH genes of *Drosophila*

The *D. melanogaster* gene *Adh*, encoding alcohol dehydrogenase, an essential enzyme for survival on medium containing ethanol, is remarkable for being transcribed from two different start-points depending on stage of development. A start-point only a few tens of bases upstream of the coding sequence functions in the later embryo up to late larval development, especially strongly in larval fat-body tissues; the second one, several hundred bases further upstream, is used during mid-embryonic development and again in the later larva and the adult. The mature mRNAs are not very different in length since most of the extra RNA transcribed from the upstream start-point is spliced out as an intron.

It is an intriguing fact that in a related species, *D. mulleri*, the two functions managed in *D. melanogaster* by one gene with two transcription starts and splicing patterns, are divided between two closely linked ADH genes, one with one mode of transcription and processing and one with the other.

A series of ingenious experiments showed that the two modes of transcription of the *D. melanogaster* gene were dependent on sequences

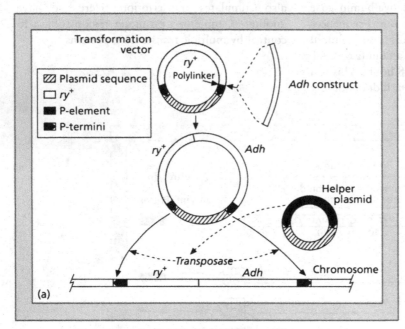

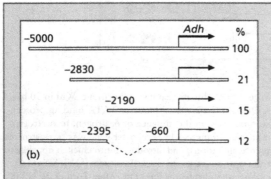

Fig. 4.15 (a) Procedure for introducing genes (in this case *Adh*, encoding alcohol dehydrogenase) into the *Drosophila* genome with a transformation vector based on the P-element (see p. 133 and Fig. 5.2). The vector contains plasmid sequence for replication in *Escherichia coli*, the *Drosophila rosy*+ (eye colour) gene flanked by P-element termini with their characteristic inverted repeats, and a polylinker sequence within which the *Adh* construct was inserted. The resulting plasmid was injected into homozygous rosy *Drosophila* embryos together with a helper plasmid containing a complete P-element to supply the transposase function needed for the excision and chromosomal integration of the *ry*+/*Adh* construct through cutting and splicing of its P-termini. Transformed flies are recognized by their wild-type (*ry*+) eye colour.

(b) Effects of deletions far upstream of the *Adh* gene on its transcription in the larva. Two segments, 660–2395 and 2830–5000 bases upstream of the promoter (shown as arrow), enhance larval transcription. Another segment, closer to the transcribed region, enhances transcription in the adult fly. After Corbin & Maniatis (1990). See also Rubin & Spradling (1982).

much farther upstream. Use was made of the transposable P-element (see p. 133), which has the ability to get itself integrated as a unit into the chromosomes. If P-DNA is injected into young embryos, some of it finds its way into cell nuclei where it becomes inserted into the chromosomes, often at multiple sites. There are two requirements for insertion: the specific sequence present at both termini of the element in reverse orientation, and a source of the transposase for catalysing the insertion. The transposase does not need to be encoded in the same piece of DNA as the one to be inserted. A transforming vector with P-element termini can be inserted into the genome by P-transposase supplied by a 'helper' plasmid injected together with the vector (Fig. 4.15a).

Through the use of this principle, variously deleted derivatives of the *Adh* gene, bracketed by P-element termini, were inserted into *Drosophila*. Selection was made initially for the *rosy*[+] marker in the transformation vector (Fig. 4.15a), and then the *Adh* construct, inserted ectopically (i.e., at an unusual locus), was transferred by crossing into a genotype in which there was no active *Adh* gene at the normal locus. RNA from various larval tissues was analysed by gel electrophoresis and specific probing for *Adh* transcripts ('Northern' blotting – compare Southern blots, see Fig. 3.1). The results (Fig. 4.15b) implicated a block of DNA extending between 660 and 5000 bases upstream of the transcription start in stimulating larval-type transcription. Deletion of either upstream or downstream segments of this 4.34-kb sequence resulted in an approximately eightfold reduction of larval *Adh* mRNA.

Thus two rather distant DNA sequences – one at least 660 bases and the other more than 3.5 kb upstream of the transcription start-point – both enhance larval *Adh* transcription. Other experiments resulted in the identification of a more closely placed enhancer sequence specific for the adult mode of transcription.

How do genes interact with their enhancers? The case of the β-globin locus control region

It is still not completely clear how enhancers work. Some of them are not just distant from but even on the 'wrong' (i.e., downstream) side of the transcription units that they control. Others (see p. 126) are situated within introns. How could such apparently inappropriately positioned sequences affect the efficiency of promoters? One likely possibility is that they make contact with promoters through looping of the intervening DNA. We saw in Chapter 1 (see p. 7) that, in metaphase chromosomes, chromatin fibres are attached in loops to a chromosome scaffold. Some of the scaffold proteins, notably the enzyme topoisomerase II, are also present in interphase nuclei, and there appear to form a more loosely organized *nuclear matrix* to which chromatin loops are still attached. Such attachment is in itself thought to be conducive to transcription. Promoters and enhancers could both be within scaffold/matrix attachment regions.

It is becoming clear that one enhancer can govern the activities of two or more genes. A prime example is the *locus control region* (LCR) lying upstream of the β-globin gene cluster of humans. The protein of haemoglobin is tetrameric, with two α-type polypeptide chains and two others, which may be ε, γ or β depending on the stage of development – ε in the embryo, γ in the later embryo and fetus, and δ and β in the minor and major components of adult haemoglobin. There are two functionally equivalent γ chains, γA and γG, differing only in a single amino acid. The genes encoding ε, γG, γA, δ and β are clustered, in that order, upstream-to-downstream (Fig. 4.16a).

The LCR was first identified as a region, between 6 and 21 kb upstream of the ε-globin gene, containing several sites that were hypersensitive to digestion by DNaseI in globin-producing tissues. DNaseI-hypersensitivity in chromatin is a consistent indicator of transcriptional activity. Its role in controlling the activity of the gene cluster was indicated by two kinds of evidence. Firstly, it was shown that one form of γδβ-thalassaemia, a rare kind of hereditary anaemia characterized by a general underactivity of the whole β-globin gene cluster, was due to a long deletion including the whole of the LCR but not the genes themselves or their individual promoters. Secondly, when trimmed-down versions of the cluster were injected as DNA into fertilized mouse eggs, it was found that the inclusion of the LCR region in the injected DNA construct greatly enhanced

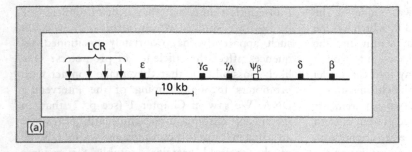

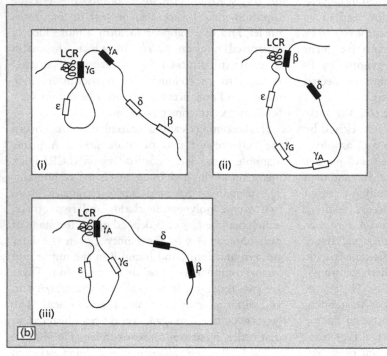

Fig. 4.16 (a) The β-globin gene cluster of humans. The vertical arrows indicate DNaseI-hypersensitive sites within the locus control region (LCR). ψ_β is a pseudo-gene (see p. 139).

(b) A hypothesis to explain competition between the promoters of the genes of the cluster. Genes are represented as boxes, black if the gene is fully capable of transcription if associated with the LCR, hatched if it has at least some potential for transcription, and open if it has been silenced by stage-specific transcriptional repressors. (i) In the fetus, ε has already been silenced and its promoter no longer competes for access to the LCR; active γ promoters, being closer to the LCR, tend to exclude δ and β. (ii) After birth, the γ genes are silenced and no longer compete; the δ and β promoters now have unimpeded access to the LCR (some additional explanation is needed for β being more active than δ). (iii) In one form of hereditary persistence of fetal haemoglobin, the γ_A promoter remains active after birth and continues to compete effectively with β and δ. Based on Grosfeld *et al.* (1993).

the expression of the human genes during the subsequent development of the transgenic mice.

The mouse experiments showed that when two genes of the cluster, β and one of the γ genes, were placed in tandem downstream of the LCR, their relative activities depended on two factors – the stage of development of the mouse and their order with respect to the LCR. It turned out that the timing of expression of the β and γ human genes was regulated in mouse in rather the same way as in humans; γ was more strongly expressed in the embryo but β came to predominate as development proceeded. Presumably this is because of the presence of different gene-specific transcriptional

regulator proteins at different stages. Superimposed on this developmental regulation, there was a marked tendency for the gene closer to the LCR to be more strongly expressed. Thus, when the order in the DNA construct was LCR-γ-β, expression of human γ was strong and expression of human β was close to zero in the transgenic mouse embryos. When the gene order was reversed, LCR-β-γ, β was expressed at quite a high level, even though γ was still somewhat stronger. These two results taken together suggest that the transcription of the β gene in the embryo was blocked when it was separated from the LCR by the γ gene. A plausible hypothesis is that the LCR enhances transcription of linked genes by making contact with their promoters, and that there is competition between promoters because only one can effectively bind to the LCR at the same time. Other things being equal, the promoter closest to the LCR will tend to block the access of the more distant one.

The hypothesis of promoter competition fits well with observations on the clinical condition *hereditary persistence of fetal haemoglobin* (HPFH), characterized by production of γ chains well beyond the fetal stage and a drastic delay in β-chain production. One mutation leading to this condition has been shown to be due to a single base-pair change in the γ_A promoter, which has the effect of abolishing a binding site for a protein called GAT1. Paradoxically, this protein was first identified as a general activator of transcription, but there are several examples of protein regulators of transcription acting as either activators or repressors depending upon the other proteins they are associated with. It is a reasonable inference that, in the context of the γ_A promoter, GAT1 acts as a transcriptional repressor, and that the loss of its binding site allows γ_A transcription to be prolonged. Whatever the mechanism, continued γ activity has the effect of depressing transcription of the β gene. This is consistent with the model diagrammed in Fig. 4.16b, in which the active γ promoter competes effectively with the β gene promoter for access to the LCR, but the inactive γ promoter does not. The suggestion is that the LCR helps the assembly of the transcription initiation complex at the promoter and, reciprocally, the proteins of that complex help stabilize the contact between promoter and LCR.

The mechanism pictured in Fig. 4.16b, with relatively distant DNA elements brought together by looping-out of the intervening DNA, recalls the idea, mentioned above, of looped attachment of chromatin fibres to a nuclear matrix. The connection between the two models remains to be established, however.

When, as in the case of the β-globin gene cluster, several genes are serviced by a common enhancer, it becomes difficult to define a single gene as including all its *cis*-acting control sequences – some of these may belong to other genes too. Evidently another term is needed to take into account multigene units subject to some degree of common *cis*-regulation. In that sense, such units are eukaryotic analogues of the bacterial operon, but they are integrated in a quite different way.

Do, in any case, enhancers necessarily always have to act in *cis*? This question is addressed in the next section.

Locus-specific *trans*-acting control – transvection

As we have seen, the concept of the gene as a unit of function becomes more complex as we take into account *cis*-acting regions that control transcription as well as the transcribed sequence itself. But at least, it is usually possible to assume that the control regions act only in *cis*. This assumption is, however, called into question by the *transvection* effect shown by several genes in *Drosophila*. This phenomenon can be best explained as due to an enhancer on one chromosome acting in *trans* to promote the transcription of a closely linked gene on the homologue.

The best-known example of transvection concerns the *Ubx* gene in the bithorax-complex (see p. 201). The defining *Ubx*[1] mutation, falling at the upstream end of the first exon, appears to nullify the gene function altogether. It is lethal when homozygous and, in heterozygotes, causes some enlargement of the posterior half of the halteres, the vestigial wings of the posterior thoracic segment (metathorax). The recessive mutations designated *bithorax* (*bx*) cause, in homozygotes, a more pronounced conversion of the halteres towards wings, and also a more general conversion of the metathorax to the form and bristle pattern of the middle thoracic segment (mesothorax). Sequence analysis shows that the *bx* mutations fall not within the *Ubx* coding region but rather within one of its long introns. They are thought to disrupt a region – a segment-specific enhancer – that is necessary for expression of *Ubx*+ in the posterior metathorax (see p. 202).

The surprising observation, made in the 1940s by E.B. Lewis, was that the *trans* heterozygote *bx* +/+ *Ubx*, was in some respects considerably more normal than the *bx* +/*bx* + homozygote. It is as if, in spite of its own failure to make an effective transcript, the *Ubx* allele is somehow helping its *bx* homologue towards normal expression. Lewis discovered that this assistance-in-

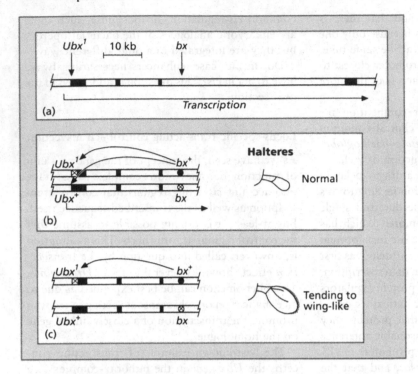

Fig. 4.17 The *Ubx* gene of *Drosophila melanogaster*, and the ability of regulatory and coding regions to partially complement one another in *trans* (transvection effect). (a) Exon/intron map of the *Ubx* gene, showing the direction of transcription and the positions of *Ubx*-null (*Ubx*[1]) and *bx* mutations. (b) Hypothesis to explain transvection. Close pairing of homologous chromosomes in the *Ubx bx*[+]/*Ubx*[+] *bx* heterozygote permits the *bx*[+] enhancer region to activate the *Ubx*[+] promoter in *trans*. The haltere, pictured to the right, is almost wild type. (c) Disruption of pairing due to a structural rearrangement takes the *Ubx*[+] promoter out of range of the *bx*[+] enhancer. The posterior part of the haltere is transformed towards wing. For further evidence, see Martinez-Laborda *et al.* (1992).

trans (transvection) was dependent on the ability of the homologous loci to undergo somatic pairing, as seen in the polytene chromosomes. Chromosomal rearrangements involving breaks close to the *Ubx* locus, making close pairing of the *Ubx* alleles difficult, made the mutant phenotype of *bx* +/+ *Ubx* heterozygotes much more extreme (Fig. 4.17).

The most likely explanation of transvection is that, when homologous chromosomes are closely synapsed, the same kind of associations that can occur, perhaps by DNA looping between enhancers and promoters on the same chromosome, can be achieved, albeit less efficiently, between closely paired homologues.

Much attention is currently focused on a gene called *zeste* (*z*), certain mutants of which suppress transvection at *Ubx* and other loci. The protein product of *z* has been shown to bind in co-operative fashion to many different loci on the polytene chromosomes, including some known to be involved in transvection effects. It seems likely that the *z*[+] product stabilizes the homologous chromosome contacts that make transvection possible.

It may well be that transvection is only signifi-

cant in organisms such as *Drosophila* where, at least in some tissues, homologous chromosomes are closely paired. Nevertheless, it illustrates a point of general significance, namely that functional contacts between chromosome sites are not necessarily absolutely dependent on their close physical linkage.

Restructured genes and edited messages

No consideration has been given in this chapter to a range of systems in which the genomic DNA is itself subject to controlled rearrangements. By far the most important from the human point of view is the vertebrate immune system, in which the great variety of cell types responding to different antigens is generated by a large number of alternative modes of DNA splicing in the differentiating cells. The genes themselves are put together from a repertoire of modules as development proceeds. This seems to be an altogether exceptional situation so far as multicellular eukaryotes are concerned, and for that reason has been put on one side in this book.

In *S. cerevisiae*, the process of mating-type

switching (see Fig. 1.5) involves the transposition of either of two alternative blocks of sequence, held permanently in reserve at loci where they are not transcribed, to the transcriptionally active mating-type locus. The fission yeast, *Schizosaccharomyces pombe*, has a rather similar switching mechanism.

The mating-type switch is the only known system of controlled DNA restructuring in the yeasts, but in unicellular animals (Protozoa), such systems are not at all unusual. The Ciliates, to cite the best-investigated group, keep their long-term genome in a *micronucleus*, and delegate everyday cell growth and proliferation to a much larger *macronucleus*, which contains a trimmed-down, shuffled and highly amplified selection of the micronuclear DNA sequences. Protozoan genetic systems, so far as they have been investigated, have been full of surprises and give the impression of being only remotely related to that of 'regular' eukaryotes.

In bacteria and bacteriophages, controlled DNA rearrangements on a smaller scale are employed fairly commonly as means of switching gene function into one path or another. The prokaryotic rearrangements all seem to be reversible.

For general reviews of programmed DNA rearrangements, see the final reference on p. 128.

Another challenging phenomenon, again set aside in this book as outside the eukaryotic mainstream, is *RNA editing*. This occurs in mitochondria of certain Protozoa and at least some plants (see Cattaneo, 1991), where the primary transcripts of some genes have their sequences altered by a form of selective copying from short 'editing' RNA sequences transcribed from elsewhere.

Conclusion – what is the gene?

Clearly, the chromosome is segmented into integrated units of specialized function, which we shall continue to call genes. But genes are not all organized in the same way, nor are they all as separate and localized as they appeared in the picture presented by classical genetics. The molecular revolution has brought both higher definition and greater complexity.

If we take into account the often very extensive flanking sequence required for normal expression and regulation of transcription, the boundaries of functional genomic units tend to become fuzzy. One way of defining the gene, which might seem satisfactory in principle though laborious to apply in practice, is as the minimum chromosomal unit which can be removed from its usual chromosomal locus and inserted elsewhere without loss of its normal function. That, however, would depend on the stringency of the test for normality. There are now many examples of transcription units which will function to some extent with very limited specific flanking sequence, but requiring distant enhancers for full and properly regulated levels of expression. Beyond enhancers, normal gene expression is dependent on features of large-scale chromosome architecture, the most obvious of which is the division into eu- and heterochromatin. We shall consider this aspect of genetic regulation in Chapter 7.

References

Almeida, J., Carpenter, R., Robbins, T.P., Martin, C. & Coen, E.S. (1989) Genetic interactions underlying flower colour patterns in *Antirrhinum majus*. *Genes Dev*, 3, 1758–67.

Catcheside, D.G. & Overton, A. (1959) Complementation between alleles in heterokaryons. *Cold Spring Harbor Symp Quant Biol*, 23, 137–40.

Cattaneo, R. (1991) Different types of RNA editing. *Ann Rev Genetics*, 25, 71–88.

Corbin, V. & Maniatis, T. (1990) Identification of *cis*-regulatory elements required for larval expression of the *Drosophila melanogaster* alcohol dehydrogenase gene. *Genetics*, 124, 637–46.

Cunningham, B.A., Himperley, J.J., Murray, B.A. *et al.* (1987) Neural cell adhesion molecule: structure, immunoglobulin-like domains, cell surface modulation and alternative RNA splicing. *Science*, 236, 799–806.

Denis-Duphil, M. & Kaplan, J.G. (1976) Fine structure of the URA2 locus in *Saccharomyces pombe*. II. Meiotic and mitotic mapping studies. *Molec Gen Genet*, 145, 259–71.

Dugaiczyk, A., Woo, S.L.C., Colbert, D.A. *et al.* (1979) The ovalbumin gene: cloning and molecular organization of the entire natural gene. *Proc Natl Acad Sci USA*, 76, 2253–7.

Fincham, J.R.S. (1966) *Genetic Complementation*. W.A. Benjamin, New York.

Freund, J.N., Vengis, W., Schedl, P. & Jarry, B.P. (1986) Molecular organization of the *rudimentary* gene of *Drosophila melanogaster*. *J Mol Biol*, 189, 25–36.

Goldberg, M.E. (1969) Tertiary structure of *Escherichia coli* β-D-galactosidase. *J Mol Biol*, **46**, 441–6.

Grosfeld, F., Antoniou, M., Berry, M. *et al.* (1993) The regulation of human globin gene switching. *Phil Trans Roy Soc B*, **339**, 183–91.

Henikoff, S., Keene, M.A., Fechtel, K. & Fristrom, J.W. (1986) Gene within a gene: nested *Drosophila* genes encode unrelated protein on opposite DNA strands. *Cell*, **44**, 33–42.

Hynes, M.J. & Davis, M.A. (1986) The *amdS* gene of *Aspergillus nidulans*: control by multiple regulatory signals. *BioEssays*, **5**, 123–8.

Jeffreys, A.J. & Flavell, R.A. (1977) The rabbit β-globin gene contains a large insert in the coding sequence. *Cell*, **12**, 1097–108.

Martinez-Laborda, A., Gonzalez-Reyes, A. & Morata, G. (1992) *Trans* regulation in the *Ultrabithorax* gene of *Drosophila*: alterations in the promoter enhance transvection. *EMBO J*, **11**, 3545–62.

Neuhard, J. & Nygaard, P. (1987) Purines and pyrimidines, In Neihardt, F. (ed.) *Escherichia coli and Salmonella typhimurium: cellular and molecular biology*, pp. 445–73. American Society for Microbiology, Washington, DC.

Newton, W.A., Beckwith, J.R., Zipser, D. & Brenner, S. (1965) Nonsense mutants and polarity in the *Lac* operon of *Escherichia coli*. *J Mol Biol*, **14**, 290–6.

Norris, R., Norris, F., Christiansen, L. & Fil, N. (1983) Efficient site-directed mutagenesis by simultaneous use of two primers. *Nucleic Acids Res*, **11**, 5103–12.

Rubin, G.M. & Spradling, A.C. (1982) Genetic transformation of *Drosophila* with transposable element vectors. *Science*, **218**, 348–53.

Ruby, S.W. & Abelson, J. (1991) Pre-mRNA splicing in yeast. *Trends Genet*, **7**, 79–85.

Souciet, J.L., Nagy, M., Le Gouer, M., Lacroute, F. & Potier, S. (1989) Organization of the yeast *URA2* gene: identification of a defective dihydroorotase-like domain in the multifunctional carbamoyl-phosphate synthetase–aspartate transcarbamylase complex. *Gene*, **79**, 59–70.

Tora, L., Gaub, M.-P., Mader, S. *et al.* (1988) Cell-specific activity of a GGTCA half-palindromic oestrogen-responsive element in the chicken ovalbumin gene promoter. *EMBO J*, **7**, 3771–8.

Varmus, H. & Brown, P. (1989) *Retroviruses*. In Berg, P.E. & Howe, M.H. (eds) *Mobile DNA*, pp. 53–109. American Society for Microbiology, Washington, DC.

Trends in Genetics 8, no. 12 (December 1992) Special issue on *Programmed DNA rearrangements*.

5

ANALYSIS OF THE WHOLE GENOME

Total DNA – quantities and patterns of complexity

C-values

The total quantity of DNA in the unreplicated haploid genome is often called the C-value of the organism. C-values for a range of species are presented in Table 5.1. There is a 300-fold difference between the smallest eukaryote genome listed (yeast) and the largest (maize), and another selection of examples could have shown differences of more than 1000-fold. Unsurprisingly the species with the least morphological complexity (yeast) has the least DNA, but, in general, the correlation between C-value and the apparent degree of evolutionary advancement is not good.

Within mammals, a rather uniform group in this respect, man and mouse have virtually the same C-value, but within the amphibia and the flowering plants there are enormous differences. Indeed, newts and lilies, near the top of their respective C-value leagues, have far more DNA than mammals. These somewhat unexpected findings are sometimes referred to as the 'C-value paradox'.

It is difficult to believe that all the DNA in the high C-value species can consist of genes with unique and essential functions, and far easier to imagine that much of it is redundant or functionless.

Degrees of repetition among genomic sequences

The first to analyse genomic DNA in terms of unique versus repetitive sequence were Roy Britten

Introduction

The preceding chapters have dealt with ways of identifying and cloning individually interesting genes and exploring their structures and functions. In this chapter we look at the genome as a whole.

The total genome can be explored at four different levels: (i) as a heterogeneous mass of DNA containing certain major components; (ii) as a set of visible chromosomes; (iii) as ever more detailed linkage maps correlated with chromosomes; and (iv) as DNA sequence.

The greatest efforts in recent years have been directed towards the development of methods for analysis of the human genome. This chapter will deal with genomic analysis of a range of organisms, but with considerable emphasis on human examples.

129

Table 5.1 Some representative C-values in terms of kilobase-pairs per haploid genome

Species	Organism	kb
Bacteriophage T4	Bacterial virus	2×10^2
Escherichia coli	Bacterium	4.5×10^3
Saccharomyces cerevisiae	Yeast	1.8×10^4
Zea mays	Maize	5.4×10^6
Arabidopsis thaliana	Thale cress	c.1×10^5
Caenorhabditis elegans	Nematode worm	c.1×10^5
Drosophila melanogaster	Fruit fly	1.7×10^5
Mus musculus	Mouse	c.3×10^6
Homo sapiens	Human	c.3×10^6

and his colleagues. They assessed the degree of repetitiveness of different components of the genome by the different rates at which they re-formed double-stranded structure after having been melted to single strands. The degree of double-strandedness at any point could be determined by passing the partially reannealed sample through a column of hydroxyapatite, a material that binds double-stranded DNA but lets single strands through. The outcome of one of these experiments, on bovine DNA, is summarized in Fig. 5.1e.

Theory (as summarized briefly in Box 5.1) predicted that the number of copies of a given DNA sequence in the genome should be inversely proportional to the *Cot value* – that is, the product of the inital concentration (C_o) of DNA in the reannealing sample and the time (t) taken for the sequence to become 50% double-stranded. To put the relationship the other way round, the complexity of a DNA sample, defined as the length of the repeating unit, is directly proportional to its Cot value.

When this kind of experiment is done with bacteriophage or bacterial DNA, the whole sample reanneals almost in unison, with a high Cot value corresponding to a complexity approximating to the size of the total genome (Fig. 5.1b,c). This is explicable on the assumption, now confirmed by very extensive sequence determination, that viral and bacterial genomes consist predominantly of unique (single-copy) DNA with little repetitive sequence. The Cot curves obtained with bacterial and viral DNAs provide reference points in the interpretation of Cot curves of larger genomes.

The result shown in Fig. 5.1e is, in fact, rather typical of what is found in eukaryotes generally. Several per cent of the DNA reassociates almost instantaneously, indicating that it is a very highly repetitive 'simple sequence'. Another 40% or so appears moderately repetitive, reannealing over a broad range of Cot values. Only a little more than a half has a Cot value of the order one would expect of single-copy DNA. In species with the highest C-values this proportion can be much less than 50%.

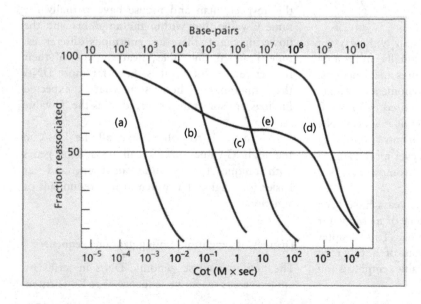

Fig. 5.1 Kinetics of reannealing of different DNA samples: (a) mouse satellite; (b) bacteriophage T4; (c) *Escherichia coli*; and (d) the separated 'single-copy' component of (e) total bovine DNA. The upper scale shows the complexity, in base-pairs, corresponding to the Cot values at 50% reannealing. Both Cot and DNA complexity are plotted on a logarithmic scale. Curves redrawn from Britton & Kohne (1968).

Box 5.1 The theory of Cot analysis

If c is the concentration of single-strand DNA in terms of number of copies (moles per litre), the rate of reassociation (loss of single strand) will be proportional to c^2.

$$dc/dt = -kc^2,$$

where k is a constant.
Integrating:

$$\int_c^{c_0} [1/c^2]dc = -k[t]dt$$

$$(1/c_0 - 1/c) = -kt$$

where c_0 is the initial concentration.
When $c = c_0/2$ (i.e., DNA 50% reassociated),

$$1/c_0 = kt_{1/2} \quad \text{and} \quad c_0t_{1/2} = 1/k$$

where $t_{1/2}$ is time for 50% reannealing and $c_0t_{1/2} = 1/k$.

If C_0 is the initial DNA concentration expressed in terms of mass of DNA per unit volume, and m is the molecular weight of the sequence reannealing,

$$C_0 = mc_0 \quad \text{and} \quad C_0t_{1/2} = mc_0t_{1/2} = m\frac{1}{k}$$

Hence $C_0t_{1/2}$ (usually written as Cot) is directly proportional to m, and also to sequence complexity in terms of base pairs.

By carrying out Cot analysis of DNA sheared to fragments of different average sizes it is possible to find out something about the interspersion of repetitive with single-copy sequences. If a single-copy sequence is present in the same fragment as a repetitive sequence it will be counted in the analysis as repetitive, because a partially reannealed fragment is retained on the separating column along with fully reannealed fragments. So if the apparent amount of a repetitive fraction is less at smaller fragment sizes it may be concluded that at least some of it is closely associated with single-copy sequences. From this kind of analysis it was concluded that, in many organisms such as *Drosophila* and vertebrates, much of the most highly repetitive sequence occurs in big blocks

However, some of it appeared to be more widely dispersed over the genome.

There is a tendency for species with higher C-values to have higher proportions of repetitive DNA. This goes some way to resolving the C-value paradox but it is not the complete answer, since such species tend to have larger amounts of single-copy DNA as well.

The C-values of different species bear no consistent relationship to the length of the chromosomes measured in recombination map units (centimorgans). The physical equivalent of the centimorgan is between one and two megabase-pairs in humans and about two kilobase pairs in yeast – nearly a thousandfold difference.

The nature of repetitive DNA

Satellite DNA

The term satellite was originally applied to DNA fractions, found in several different vertebrate animals, that were separable from the bulk of the DNA by buoyant density. When a solution of caesium chloride is centrifuged at very high speed, the heavy caesium ions tend to sediment to the bottom of the tube, so establishing a density gradient, heavy at the bottom of the tube and lighter at the top. If a DNA sample, sheared to fragments of a few kilobases, is added to the tube before centrifuging, each DNA fragment will come to equilibrium as a band at the position in the gradient corresponding to its own buoyant density. Density depends upon base composition, G–C being somewhat denser than A–T base-pairs. A satellite fraction is recognized as a minor band separated by density from the band formed by the bulk of the DNA. It consists of relatively simple and highly repetitive sequences that happen to differ from the bulk DNA in base composition. The term satellite tends now to be used to describe all tandemly repetitive high copy-number DNA sequences, even though not all are separable by density. Such sequences account for the highly repetitive non-dispersed DNA fraction identified by Cot analysis.

Satellite DNA has been most intensively investigated in *Drosophila* species, mice, rats and humans. In all these organisms it comprises a substantial fraction of total DNA – typically in the range 5–10%. It is a major component of heterochromatin, defined as those parts of the chromosomes that remain condensed and deeply-staining throughout the cell cycle, in contrast to the remainder, euchromatin, which is extended and diffuse in non-dividing nuclei. Satellite is typically arranged in blocks of thousands or millions of tandem repeats with minor variations, often concentrated around centromere regions of chromosomes. In *Drosophila melanogaster* 10 different repeat units have been identified, all of them A/T-rich. For example, two of the simplest and most abundant (3 and 6% of the genome, respectively) have the sequences 5'AATAT3' and 5'AAGAG3'. Other satellites are more complex; in the mouse the major repeated unit, which accounts for about 10% of the genome, is 234 bp long, with less precise internal tandem repetition.

In humans, in addition to a substantial amount of heterochromatin-associated satellite DNA, there are other less abundant simple sequences which occur scattered over the chromosomes in small arrays of tandem repeats, the numbers in each array often varying from one individual to another. One such dispersed simple sequence, called minisatellite, has been important in forensic medicine. Southern analysis (cf. p. 69) of human DNA using a minisatellite probe gives a unique pattern of hybridizing bands for each individual – a DNA 'fingerprint'.

Satellite DNA has no identified function for the organism. The repetitive sequences of which it is composed vary greatly between different species, even, as in *Drosophila* and mice, within the same genus. But, within a species, the same satellite sequences often recur on every chromosome. It looks as if different repeats have been independently proliferated in each species – perhaps by cycles of excision, amplification and reinsertion into the chromosomes, especially into centromere regions.

Functional genes in repetitive arrays

Another highly repetitive component of all eukaryotic genomes is the ribosomal RNA (rRNA) gene complex, the internal structure of which was described in the last chapter (see Fig. 4.8). rRNA genes (called rDNA) are typically clustered in tandem arrays at one or a few chromosome loci, frequently identifiable microscopically as nucleolus organizers. The *nucleolus* is a spherical mass of nascent ribosomal material which is usually prominent in non-dividing and prophase nuclei. Nucleolus organizers can be recognized by their attachment to the nucleolus in prophase and as thin, non-condensed segments at metaphase and anaphase. They can also be heavily labelled by rDNA probes. Most species have a few hundred rRNA gene copies; the estimates for humans, *Drosophila* and yeast are 200, 200 and 140, respectively. This means that rDNA is a fairly major component (about 7%) of the total yeast genome, where it is by some margin the most abundant repetitive sequence, but contributes proportionately much less (about 0.3%) to the bulk of the human genome. In addition to the DNA sequences transcribed to give the 18S-5.8S-18S precursor (see Fig. 4.8), all eukaryotes have multiple genes specifying an additional small (5S) rRNA. In yeast, these genes form part of the major rDNA cluster but in multicellular eukaryotes they form separate clusters, amounting in all to several hundred copies.

rRNAs are required in large amounts in all growing cells as essential components of the protein-synthesizing machinery. That, presumably, is why the genes specifying them need to be present in such large numbers. Histones, the major chromosomal proteins, are also in heavy demand, at least in dividing cells. So it is not surprising to find that the genes encoding the five classes of histones are also highly reiterated. In all species where their arrangement has been characterized, histone genes occur in clusters, one gene of each kind in each cluster, not necessarily all oriented in the same direction. The clusters themselves tend to be clustered. In *D. melanogaster*, for example, there is a closely linked array of about 100 repeats. In humans the number is approximately 20. rRNA genes, histone genes and other reiterated protein-encoding genes which are less universally distributed, together account for a substantial proportion of the middle-repetitive fraction of the genome.

Dispersed repeats

One of the major revelations in genetics over the past two decades has been the extent to which genomes have been infiltrated by elements that seem to be present simply because of their capacity for proliferation and dispersal. Some of these apparent 'DNA weeds' encode at least a part of the apparatus needed for their own proliferation, while others (the pseudogenes and SINES, reviewed below) seem to have been spread willy-nilly by agencies outside their own control. Some appear to be transposed around the genome by excision and reinsertion, but many others move via RNA intermediates. We shall consider the different kinds of movable element in turn.

Transposons

Transposons were first identified in bacterial plasmids as movable elements carrying genes for antibiotic resistance. The usage of the word is now generally extended to include eukaryotic DNA elements of the kind first identified by Barbara McClintock in maize (see Fedoroff, 1988), where they are the cause of unstable mutations – losses and restorations of gene activity as the elements are inserted into and excised from gene loci. One such element in another plant, *Tam3* of *Antirrhinum majus*, was introduced in Chapter 3 in the context of gene tagging (see Fig. 3.14).

McClintock identified two families of transposable elements in maize (one, the *Ac–Ds* family, is shown in Fig. 5.2), and several others are now known. Within each family, some members, like *Ac*, are able to transpose autonomously and others, like *Ds*, transpose only in response to the activity of an autonomous element of the same family located elsewhere in the genome. Autonomous transposition depends on the presence of an open reading frame that is presumed to encode an enzyme function (transposase) required for excision and reinsertion. The ability to transpose, whether autonomously or not, depends on a special sequence – a cutting site – recognized by the transposase and present at the ends of the inverted terminal repeats that are a defining feature of transposons in general. The dependent elements of a family are evidently derived from their autonomous relative by loss, generally by deletion, of an essential part of the open reading frame; they retain the inverted repeats that make them targets for the transposase.

The best-investigated *Drosophila* transposon is the P-element (see Engels, 1989), one of two known alternative causes of *hybrid dysgenesis*, the other being the LINE element I – see below. P shares key features with the plant transposons, namely a long open reading frame encoding a transposase, and terminal inverted repeats on which the transposase acts (Fig. 5.2). Again like the plant elements, it gives rise to numerous non-autonomous derivatives. The hybrid dysgenesis syndrome – chromosome breakage and gene disruption – is a consequence of hyperactive P-element transposition in the progeny of crosses

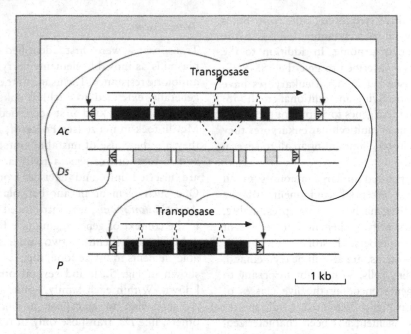

Fig. 5.2 Structures of some transposons from eukaryotes. *Ac*, Activator element of *Zea mays*: the open reading frame (ORF), interrupted by introns, is represented by the filled boxes, the unit of transcription by the arrow, and the terminal inverted repeats by hatched arrowheads. The processed transcript encodes a transposase protein that can cleave the element at its termini and facilitate its transposition to another chromosomal locus. The *Ds* family of elements, derived from *Ac* by deletion in the example shown, have *Ac* terminal inverted repeats but lack the complete ORF and so cannot produce transposase. They can be transpose if there is an intact *Ac* elsewhere in the maize genome. The *Drosophila melanogaster* P-element, like *Ac*, encodes a transposase that acts on its own inverted-repeat termini. The third intron is spliced out only in the germline, and so P does not normally transpose in somatic cells.

between females lacking the element and males possessing it. For reasons that are still in dispute, P-elements tend to become quiescent in *Drosophila* strains that have harboured them over a number of generations, but this acquired immunity is transmitted only through the female.

A peculiarity of the P-element is that the third intron interrupting its open reading frame is spliced out only in the germ-line, not in somatic cells. Hence the effects of P-transposition are normally seen only in the next generation. To make a P-element behave like a plant transposon and cause somatic variegation, it is necessary to remove its third intron precisely by DNA manipulation *in vitro*. We shall return to tissue-specific intron splicing in the context of normal developmental control in Chapter 7 (see p. 192).

Transposable DNA sequences may increase their copy number in more than one way. They can, as demonstrated for some of the maize elements,

transpose from a locus already replicated to another just about to replicate, thus losing one copy but getting two back. In *Drosophila* hybrid dysgenesis, the P-element copy number can build up so quickly that it is suspected that replication can occur outside the chromosome between excision and reinsertion.

Transposons have not been reported in human or other mammalian genomes, and seem not to be major DNA components in any species. Even *Drosophila* P-elements, which can build up to relatively high copy number, hardly exceed 0.1% of total DNA. Their importance, in the species in which they occur, is as agents of mutation and chromosomal rearrangement.

Retrotransposons

Retrotransposons differ most conspicuously from transposons in being flanked by long terminal

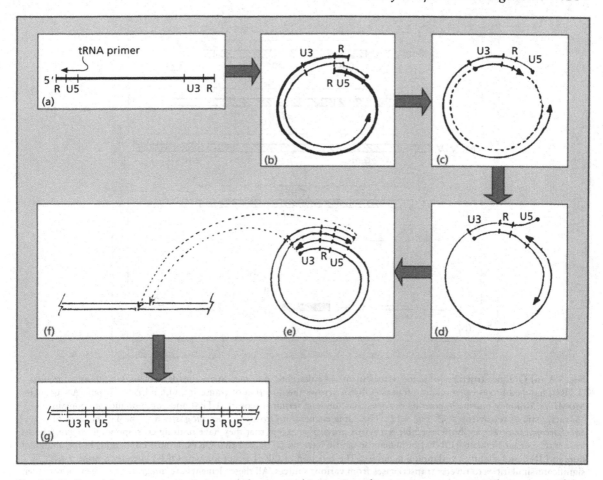

Fig. 5.3 Outline of the reverse transcription and chromosomal integration of a retrovirus genome. (a) The structure of the infective viral RNA with tandemly repeated (R), 5' and 3' unrepeated (U5, U3) segments, and 3' poly-A tail (An). After infection into the host cell, reverse transcription is primed by a host transfer RNA molecule which anneals to the viral sequence just 3' to U5. (b) The reverse transcript, having passed through U5 and part of R, dissociates from the 5' R segment, reanneals to the 3' R segment and continues anticlockwise through U3 to the rest of the viral RNA. (c) The already-copied U3-R RNA is digested back to the 5' border of U3, and primes second-strand DNA synthesis clockwise through U3 and R. After copying, all the RNA is degraded. (d) The first DNA growing point, having completed synthesis up to U5, makes a second 'jump' and, using the second strand as template, proceeds through U5, R and U3, displacing its own 3' end. (e) The first and second DNA strands copy each other to their respective 5' ends; the result is a complete double-stranded DNA copy of the viral genome with tandemly repeated U3-R-U5 termini. This copy is ligated into a staggered double-strand chromosome break (f) to become an integrated provirus (g). The gaps due to the single-strand overhangs on each side of the chromosome break are filled in by DNA polymerase to create short tandem host-site repeats flanking the provirus. DNA and RNA strands are represented by thin and thick lines respectively, and their ends are distinguished by arrowheads (3', potential growing-points) and dots (5'). Based on Varmus & Brown (1989).

direct repeats (LTRs). They appear to move not by excision but rather by transcription, reverse transcription and chromosomal integration of the reverse-transcripts. In all these characteristics they resemble chromosomally integrated retroviral genomes (*proviruses*). The life cycle of a typical retrovirus is sketched in Fig. 5.3. As already shown in Fig. 4.11, the retroviral genome encodes reverse transcriptase as well as the proteins of the infective virus particle.

Retrovirus-related elements have been found in the genomes of a wide range of organisms. In

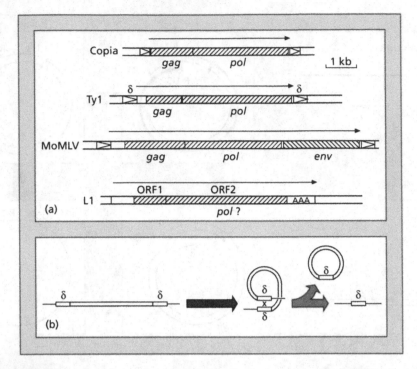

Fig. 5.4 (a) Outline structures of some retroelements of eukaryotic genomes. Arrowheads indicate long terminal repeats (LTRs), hatched boxes open reading frames (ORFs), arrows the direction of transcription, and AAA 3′ poly-A sequences. MoMLV, Moloney murine leukaemia provirus, has the long terminal repeats (LTRs) and *gag, pol* and *env* domains characteristic of retroviruses (cf. Fig. 4.12). The copia element of *Drosophila melanogaster* and the Ty1 element of *Saccharomyces cerevisiae* both resemble retrovirus proviruses except that they have nothing corresponding to *env*. The L1 long interspersed elements (LINEs) of human or mouse have no LTRs and two ORFs, almost contiguous in the abundant human LINE and slightly overlapping in mouse. The function of ORF1 is not known; ORF2 encodes a protein with significant similarities to reverse transcriptases from various sources. All these elements are integrated between short tandem repeats generated from the target site in the chromosome (see Fig. 5.3). (b) How a Ty element can be excised from the chromosome by recombination between its terminal δ repeats, leaving one δ behind.

mammals, particularly in mouse, proviruses, usually decayed by mutation to functional ineffectiveness, have been identified as insertions in mutant genes, where they were presumably the original cause of mutation. Defective proviruses are fairly common in vertebrate genomes, but generally appear to be totally quiescent.

In *Drosophila* and *Saccharomyces*, on the other hand, elements with long direct terminal repeats and other proviral features are actively transcribed and transpose, through reverse transcription, at some low frequency (see Boeke, 1989; Bingham & Zachar, 1989). Unlike the transposons decribed in the last section, they get inserted at new loci without removal from their previous sites of integration.

Of the several families of yeast retrotransposons, the most abundant are the quite closely related Ty1 and Ty2 elements, which are typically present in the yeast genome in about 30 and 10 copies, respectively. Together they constitute perhaps 0.1% of the genome. In addition to the full-length Ty elements, there are usually of the order of 100 isolated copies of their LTRs (called δ). These are presumably derived from complete Ty elements by recombination between a site on one LTR and the corresponding site on the other, resulting in deletion of the sequence in between (Fig. 5.4b). Yeast cells are hyperactive in homologous recombination generally, and so direct repeats in the yeast genome are prone to this sort of accident.

In *Drosophila* retrotransposons occur in astonishing profusion. At least 10 kinds have been characterized, and the tally seems constantly to increase. At 10–80 copies each (see Fig. 5.8), they comprise at least 2% and perhaps something closer to 10% of the total genome.

The structures of two retroposons, Ty1 from yeast and *copia* from *Drosophila*, are shown in Fig. 5.4. They differ from a typical retroviral provirus in lacking the *env* domain. When transcribed at artificially high levels, both these retrotransposons have been found to make intracellular virus-like particles, presumably lacking only the *env*-encoded components necessary for cell-to-cell infection.

As compared with some of the plant transposons or the *Drosophila* P-element in its active phase, the yeast and *Drosophila* retrotransposons move rather rarely. Nevertheless, many of the spontaneous mutations that have been characterized at the DNA level in these species turn out to be due to retrotransposon insertions. This is true of a high proportion of the mutants used in classical *Drosophila* genetics.

Long interspersed elements (LINEs)

LINEs occur widely in eukaryotes, but are especially prominent in some mammals. The mouse L1 element (Fig. 5.4) is a good example. At its full length it measures 6 kb, but occurs in a range of forms shortened from the 5' end. It contains two open reading frames overlapping by 14 bases. The upstream open reading frame is fairly short (1137 bp) and does not encode anything recognizable; the downstream one (3900 bp) encodes a protein with strong similarity in several parts of its sequence to reverse transcriptases from various sources. At the 3' end the element terminates in a 'tail' consisting predominantly of adenylate residues. These features, taken together, suggest that the element has been spread through the genome as a polymerase II (PolII) transcript, polyadenylated at the 3' end and reverse-transcribed by an enzyme encoded in its own sequence. The large number of 5'-truncated elements can be explained as incomplete reverse transcripts, all initiated in the 3' poly-A tail but not reaching the 5' end.

The sheer bulk of the L1 component of the mouse genome is remarkable (see Fig. 5.15). There are up to 100 000 copies of various lengths, averaging perhaps 2.5 kb and amounting to approximately 250 megabases or over 8% of the entire genome. It is scattered throughout the genome and is likely to be found almost anywhere where an insertion could be present without disrupting an essential function.

The human genome also contains L1 elements, roughly similar in size structure, quantity and distribution to those in mouse. Human L1s are sometimes called Kpn elements, because they typically contain a *Kpn*I restriction site. Although they may be found almost anywhere in the genome, they appear to be more abundant in those chromosome segments characterized microscopically as G-bands (see p. 140).

Other LINEs have been found in certain strains of *Drosophila*, maize and the fungus *Neurospora crassa*. The best-known *Drosophila* LINE, called the I factor, resembles the otherwise very different P-element in causing hybrid dysgenesis in the progeny of crosses in which it is contributed only by the male parent. The occasionally disruptive activity of I is not typical of LINEs in general. Fortunately, human L1 elements seem completely stabilized, though they must have gone through a highly proliferative phase at some time in the distant evolutionary past.

Short interspersed elements (SINEs)

The most famous example of a SINE is the human *Alu* element, so-called because it typically contains a centrally located *Alu*1 restriction site. It has a great deal of sequence similarity to an abundant small RNA called 7SL, which is a component of the *signal recognition particle*, a ribonucleoprotein complex which functions in the positioning of proteins in cell membranes. The *Alu* sequence of about 300 bp corresponds to two somewhat diverged 7SL sequences arranged head-to-tail. Like LINE elements, *Alu* has an A-rich sequence at its 3' end.

Alu is extraordinarily abundant in the human genome (see Fig. 5.17). It is present in about half a million copies, averaging one copy per 6 kb and accounting for about 5% of the total DNA. A specific probe for human *Alu* will hybridize to most random human genomic fragments of more than a few kilobases in length, and can be used to prime specific amplification of human DNA present in human–rodent cell hybrids (see p. 143).

Various other SINE elements occur in other mammalian species, and all appear to be related to small RNA molecules with essential cell functions. The mouse B1 element, like the human *Alu*, is related to 7SLRNA, but the SINE element found in the galago monkey is clearly related to methionine-specific transfer RNA (tRNA) and that in goats to valine-specific tRNA. What all these RNAs have in common, apart from their small size and constant presence in the cell, is their transcription by RNA polymerase III (PolIII). PolIII transcription is initiated from promoters internal to the transcribed sequence (see Fig. 4.8), a fact that provides a possible explanation for the extraordinary degree of proliferation of SINEs. If, as a consequence of the presence in the cell of reverse transcriptase, PolIII transcripts are reverse-transcribed and reinserted into the chromosomes, they will carry their internal promoters with them and will thus be available for further rounds of transcription and reverse transcription. This sets up the possibility of epidemic spread provided there is some source of reverse transcriptase. The enzyme is clearly not encoded in the SINEs themselves, but might be present from time-to-time as a consequence of retrovirus infection or LINE activity. A favoured hypothesis for the proliferation of SINEs, which also provides an explanation and a role for the A-rich 3′ termini, is presented in Fig. 5.5. A more special scenario is required for the origin of the double structure of human *Alu*.

Pseudogenes

It has been a common experience that screening a genomic library with a copy DNA (cDNA) probe can turn up sequences that, though closely related to the sought-for gene, are inactivated by frameshifts and deletions. These pseudogenes are of two kinds.

The first appears as a redundant and non-functional copy in a series of tandemly arranged functional repeats. The classical examples are in the α- and β-globin gene clusters of mammals (see Fig. 5.17b). It can be argued that these could have arisen by out-of-register crossing over between different members of the tandem array. One product of such unequal exchange will have lost a gene and will presumably be lost by natural selection; the reciprocal product will have an extra and redundant gene that can suffer mutational inactivation without detriment to the organism (Fig. 5.6a).

Pseudogenes of the second kind are not members of tandem arrays and tend to turn up in positions quite unrelated to those of their functional relatives. Their structures strongly suggest that they arose by insertion of reverse-transcripts of messenger RNAs (mRNAs). They do not include polymerase II (PolII) promoter sequences, and are therefore not transcribed. Like PolII transcripts, they have poly-A sequences at their downstream ends. Most strikingly, they have lost all introns. For this reason they are often called *processed pseudogenes*, to distinguish them from the category described in the preceding paragraph.

Although they are individually not very numerous, processed pseudogenes may well, at least in mammals, exceed the corresponding active genes in copy number. However, since they lack introns, they contribute relatively little to the total DNA.

Chromosome characterization by microscopy

Features visible in unspecifically stained chromosomes

The best general-purpose chromosome stains, such as orcein and hematoxylin, bind to chromatin indiscriminately. They enable distinctions to be made between chromosomes in so far as they differ appreciably in size or centromere position. The centromere of a metaphase chromosome is usually clearly recognizable as a constriction where the chromatids are held together. General stains also sometimes reveal other distinguishing landmarks such as *nucleolus organizers* (rDNA loci), and, at least at prophase stages of nuclear division, blocks of heterochromatin and shorter bead-like regions of high density called *chromomeres*.

The bands of *Drosophila* polytene chromo-

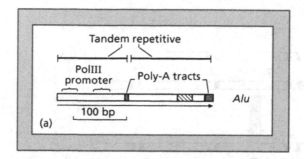

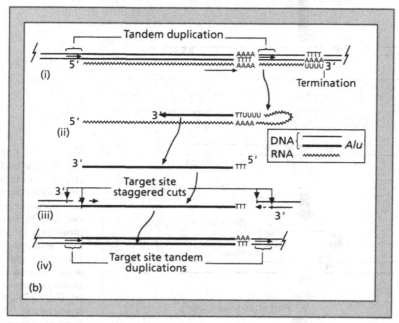

Fig. 5.5 (a) The structure of the *Alu* element, the abundant human short interspersed element (SINE). The two parts of the element show considerable sequence similarity to each other (apart from an insertion in one of them), and to the 7SL RNA of the signal recognition particle; the left-hand segment has sequences resembling a typical bipartite PolIII promoter, rightward-directed. (b) A hypothesis to explain the proliferation of SINEs. (i) The element is transcribed by reason of its PolIII promoter. The transcript over-runs the A-rich 3′ end of the SINE and terminates in a run of Us (As in the template DNA). (ii) The terminal U-rich sequence folds back and anneals to the A-rich sequence, priming reverse transcription. (iii) The reverse transcript is ligated to chromosomal DNA at a staggered break, and (iv) integrated in double-stranded form by repair-filling of the single-strand gaps and ligation of the loose ends. The integrated *Alu* copy retains its internal promoter for further transcription. The repair of the staggered break generates the short flanking target-site duplications characteristic of transposable elements in general. After Deininger (1989).

somes, described in Chapter 2 are chromomeres at a much higher level of resolution. A detailed map of the banding patterns of all four chromosomes, comprising some 5000 bands of differing intensities, is part of the stock-in-trade of the *Drosophila* geneticist. Exhaustive mutagenesis and complementation analysis (see Fig. 2.2) has led to the general conclusion that there is approximately one functionally essential gene per band, though a strict gene–band correlation cannot be relied upon. The heterochromatic segments flanking *Drosophila* centromeres, though prominent in the small metaphase chromosomes, do not become polytenized in salivary gland nuclei and are thus scarcely represented at all in the polytene chromosome maps.

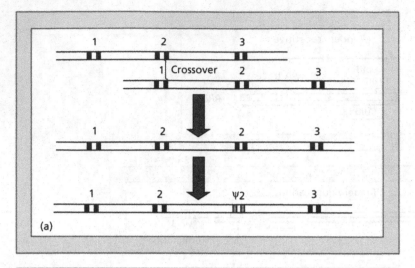

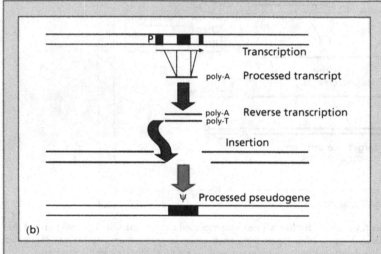

Fig. 5.6 Two origins for pseudogenes. (a) Out-of-register crossing over between genes in a tandemly repetitive array, resulting in a redundant gene copy which can decay by mutation to a pseudogene (ψ) without detriment to the organism. This scenario is particularly applicable to the origin of pseudogenes within the globin gene clusters of mammals (cf. Fig. 5.17b). (b) Chromosomal integration of DNA reverse-transcribed from a processed PolIII transcript. This kind of pseudogene lacks introns and typically has a 3′ poly-A tract.

In organisms other than *Drosophila*, it is the metaphase chromosomes that provide the most information. They may reveal less detailed structure than prophase chromosomes, but, because of their compact form, they are much easier to resolve under the microscope.

Chromosome banding

The possibility of distinguishing different metaphase chromosomes by microscopy has been greatly increased by the development of a series of special chromosome banding techniques. There are three main banding procedures with minor variations.

G-banding uses Giemsa stain, originally used in bacteriology. It involves pretreatment of the chromosomes with a warm saline solution or with the proteolytic enzyme trypsin to remove some of the protein. G-bands are regions of relative compaction of the chromatin, and correspond more or less to the chromomeres recognizable at prophase (cf. Fig. 1.1b(i)). Figure 5.7 shows an example of the use of G-banding to distinguish human metaphase chromosomes.

Q-banding is obtained with a variety of stains,

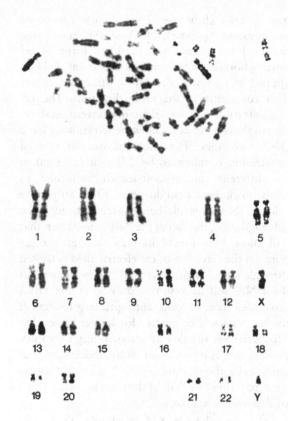

Fig. 5.7 G-banding of human male metaphase chromosomes. The upper part of the figure shows the stained chromosomes in their original orientations in the squashed metaphase nucleus, and the lower part shows the same chromosomes rearranged in homologous pairs and numbered. Courtesy of Professor H.J. Evans.

including quinacrine, that fluoresce in ultraviolet light. Quinacrine binds preferentially to DNA that is relatively rich in A–T base-pairs. On the whole Q-bands coincide with G-bands.

R-banding is so called because it gives the reverse image to G-banding. It involves Giemsa staining following treatment of the chromosome preparation with hot buffer solution that melts the more A/T-rich DNA regions to single strands. Partially single-stranded DNA binds the stain much less well than fully double-stranded DNA, and so the R-bands represent the more G/C-rich regions between the G-bands.

These banding techniques are of quite general

application but have been particularly important applied to human chromosomes. At metaphase, about 850 G-bands can be identified on the 24 kinds of chromosomes in the human genome. The coarse-scale subdivisions defined by the most prominent bands and designated according to a standard system of letters and digits provide reference points for finer-scale mapping using DNA markers (see the example of chromosome 21, Fig. 5.16).

G-, Q- and R-bands are useful visible landmarks in the human genome, and they correlate to some extent with chromatin function. G (and Q) bands have some of the characteristics of hetero-chromatin. They tend to replicate later than R-bands, are relatively poor in genes (though numerous genes have been located within them). They have less acetylation of their histones – a rather general indication of gene under-activity as we see in Chapter 7 (Fig. 7.17). The R-bands are relatively gene-rich and have more histone acety-lation. For no clear reason, the two kinds of bands also differ (in humans) in their overall content of LINEs and SINEs; R-bands have more *Alu* and G bands more L1 elements (Craig & Bickmore, 1993).

Reference has already been made (see p. 90) to the use of specific radioactive or fluorescent probes for determining the positions on chromosomes of particular DNA sequences. Through the use of different probes fluorescing in different colours – the fluorescent *in situ* hybridization (FISH) tech-nology for 'chromosome painting' – two or more sequences can be probed for simultaneously (Ried *et al.*, 1992). The precision of the fluorescent probes is limited only by the resolving power of the light microscope. This is very much better than the older method of probing by radioac-tivity (tritium), which always gives a scatter of silver grains in the radiosensitive film around the labelled locus.

An example of sequence-specific chromosome probing is shown in Fig. 5.8. Any gene whose DNA sequence is even partly known can be located within the chromosome segments defined by banding. In *Drosophila* polytene chromosomes, a cloned single-copy gene can usually be located in just one of the 5000 or so visible bands.

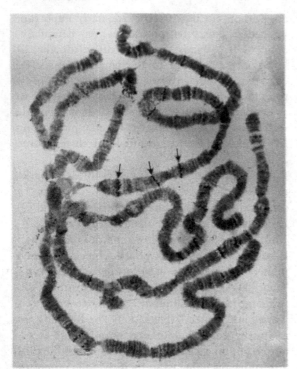

Fig. 5.8 Locating DNA sequences on chromosomes by means of sequence-specific DNA probes. Locating the multiple copies of a copia-like transposable element (see p. 137) in *Drosophila melanogaster* salivary gland polytene chromosomes, using a tritium-labelled probe. The five long chromosome arms (one for the x-chromosome, two each for chromosomes 2 and 3) radiate from a heterochromatic *chromocentre* within which all the centromeres adhere. The chromosome bands harbouring the transposable element (about 30 in all, of which three examples are arrowed) are revealed by the black silver grains in the superimposed radiosensitive film. Photograph courtesy of Professor D.J. Finnegan.

Chromosome sorting – chromosome-specific probes and libraries

Analysis of the whole genome, chromosome by chromosome, can be greatly facilitated by methods for separating out individual chromosomes or, even better, chromosome fragments. There are several possible methods that can be applied in different cases.

Pulsed-field gel electrophoresis (PFGE)

In yeast and filamentous fungi the chromosomes are so small that the size of the DNA molecule from a single chromosome is only a few hundred or thousand kilobases. DNA in this size range can be fractionated by a modified form of gel electrophoresis in which the electrical field is applied in pulses in alternating directions rather than constantly in the same direction. The gel concentration is lower than in conventional gel electrophoresis so as to allow penetration of large DNA molecules. The effect of size on rate of penetration is enhanced by different times taken for different chromosome-length molecules to adjust to changed field direction. One can picture a long DNA molecule being carried through the gel lengthwise, this being the only orientation that will allow it to thread its way through the gel pores. If the direction of the electric field is shifted through some angle – say 90° – movement of the DNA will be resumed only after the DNA molecules have become appropriately realigned; the lag time will be greater for longer molecules. This is an over-simple way of explaining a complex process that is by no means fully understood, but empirically the method works. Figure 5.9 shows the separation by PFGE of the chromosomal DNA of *Aspergillus nidulans*.

PFGE provides a way of assigning cloned genes to chromosomes that are too small for useful microscopy. The different gel bands can be blotted onto a membrane, essentially as in ordinary Southern analysis but with treatment of the gel after electrophoresis with ultraviolet light to break the separated chromosome-sized DNA bands to smaller fragments that will blot more readily. Hybridization of the blot to gene-specific probes will then reveal which genes are located in which band (Fig. 5.9).

Fluorescence-activated chromosome sorting (FACS)

Whereas PFGE fractionates purified chromosome-sized DNA molecules that have been freed from proteins, FACS fractionates intact metaphase chromosomes isolated from cell cultures that have been arrested in metaphase by a spindle-inhibiting drug, either colchicine or colcemid. The sorting is on the basis of two different criteria, size and overall base-composition. The chromosomes are stained with a mixture of two dyes that fluoresce in different colours in response to

Fig. 5.9 An example of fractionation of chromosomal DNA from a small genome by pulsed-field gel electrophoresis (PFGE). The left-hand part of the figure represents chromosomal DNA of the fungus *Aspergillus nidulans* separated on an electrophoretic gel and stained for photography with ethidium bromide. The approximate sizes in megabase-pairs (Mb) of the separated fractions are indicated. Two pairs of chromosomes are too similar in size for separation. To the right is a Southern blot of four such separations in parallel hybridized to four different radioactive gene-specific probes. It is evident that *trpC* is on chromosome 8, *wetA* (a gene involved in asexual spore formation) is on 7, *lysF* on either 1 or 5, and *argB* on either 3 or 6. After Brody & Carbon (1989).

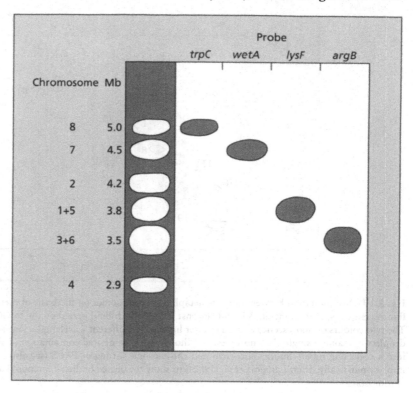

different wave-lengths, one in the visible and one in the ultraviolet range. One dye binds preferentially to AT-rich and the other to GC-rich sequences. A stream of stained chromosomes in liquid suspension is converted into a spray of droplets, each one so small that it is unlikely to contain more than one chromosome. The droplets are monitored individually by passage through laser beams tuned to the wavelengths needed to excite the fluorescence of the two dyes. The two colours of fluorescence are recorded photometrically. Each kind of chromosome gives a characteristic combination of signals; the amount of fluorescence is dependent on chromosome size and the ratio of the two colours varies according to DNA composition. The chromosomes of the human genome can all be distinguished with relatively little error (Fig. 5.10). The fluorescence detection system can be set to recognize a particular kind of chromosome and activate an electric field to deflect droplets containing that chromosome into a collecting tube.

Use of human–rodent hybrid cells

Human–rodent cell hybrids have played an essential part in the analysis of individual human chromosomes. Mammalian cells in culture can be induced to fuse, and a mixture of either human and mouse or human and hamster cells can be used to make inter-species cell hybrids. Cell fusion is usually followed by nuclear fusion, and clones of hybrid cells initially have complete diploid sets of human and rodent chromosomes together in the same nucleus. As growth proceeds, the human chromosomes tend to get lost and the rodent chromosomes retained, so that subclones can be isolated in which only one or a few human chromosomes survive (Fig. 5.11).

The identity of the surviving human chromosomes can be established by chromosome banding. They can be used, as described below, as a source of chromosome-specific probes.

The hybrid cell technique can also be used for isolation of human chromosomes with segmental

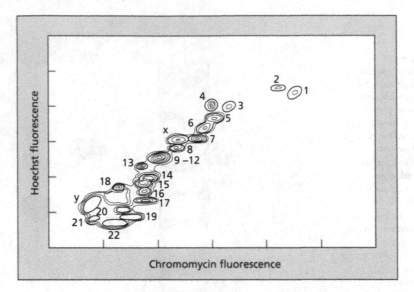

Fig. 5.10 Discrimination between human metaphase chromosomes on the basis of their differential binding to two fluorescent dyes, chromomycin A3, and Hoechst 33258 which bind preferentially to GC- and AT-rich DNA, respectively. The two colours of fluorescence excited by laser light at two different wavelengths (one for each of the dyes) are recorded for droplets containing single chromosomes, and those giving the desired combination of signals are deflected electrostatically into a collecting tube – fluorescence-activated chromosome sorting or FACS (see also text). Redrawn, slightly impressionistically, from Langlois *et al.* (1982), to show the degree of discrimination that can be achieved between different chromosomes.

rearrangements. Cells from people discovered by clinical cytogeneticists to be carrying segmental interchanges can be cultured and fused with rodent cells. Cell lines can subsequently be isolated that each contain a rearranged human chromosome as the sole human nuclear component. Such isolated chromosome rearrangements can be very useful for mapping molecular markers (see Fig. 5.16).

A modification of the same general method is used for the isolation of fragments of human chromosomes. Human cells or, better, hybrid cells with just one human chromosome are heavily irradiated with X-rays to break the chromosomes into fragments before fusion with non-irradiated rodent cells. Few if any of the fragments are able to persist autonomously, but some, presumably a random sample, are rescued by becoming integrated into the non-irradiated rodent chromosome (Fig. 5.12). The presence of even a small part of a human chromosome in a rodent genome can be detected by means of a probe specific for the human *Alu* element, even if not by chromosome banding. In different cell lines different human chromosome fragments are retained, and panels

of cell lines containing different and overlapping segments of the same chromosome can be used for mapping (*radiation hybrid mapping*, see p. 153).

Physical dissection of the chromosomes

Drosophila polytene chromosomes can be dissected physically with ultra-fine glass needles and a micromanipulator, and sections containing just one or a few bands transferred to small containers for extraction of DNA. This extraordinarily delicate technique can be applied even to human metaphase chromosomes, though here, of course, the degree of resolution is much less – perhaps a thousand genes in a dissectable piece rather than ten or fewer. The tiny amounts of DNA so obtained can be selectively amplified by the *polymerase chain reaction* (PCR) technique, described below.

Getting chromosome-specific DNA markers – the use of PCR

DNA from sorted or dissected chromosomes can in principle be used for making chromosome-

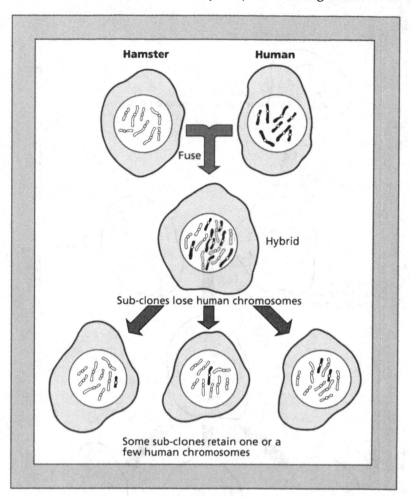

Fig. 5.11 Derivation of hybrid cell lines containing single human chromosomes. For simplicity only four pairs of chromosomes of each genome are shown.

specific DNA libraries, which can be extremely useful in the construction of detailed chromosome maps. The technical difficulties are the small amounts of DNA obtainable from chromosome sorting and, in hybrid cell preparations, the contamination of human with rodent DNA. These problems can to large extent be overcome by specific amplification of human DNA sequences by *PCR*. The principle of this technique is outlined in Fig. 5.13a (see p. 148). It requires some knowledge of DNA sequences flanking the segment that one wishes to amplify. A heat-resistant DNA polymerase isolated from a thermophilic bacterium such as *Thermophilus aquaticus* (Taq) is used to replicate single-stranded target DNA to which single-stranded flanking sequences have been annealed as inward-directed primers. Following each period of polymerization, the resulting double-stranded fragments are melted again to single

strands, which are annealed again to the primers for another round of polymerization. Ideally, each round doubles the quantity of specifically amplified DNA. The cycle is controlled by a computer programme of temperature shifts: for example, 1 minute at 92 °C for melting, half a minute at 58 °C for annealing and 2 minutes at 72 °C for polymerization, repeated perhaps 30 times. At the end of the programme a sample of the reaction is run on an electrophoretic gel and, if all has gone well, a strong band will appear, far exceeding in amount the small smample of DNA used to seed the reaction.

Everything depends on the availability of the specific primers. They need only be about 20 nucleotides long and can easily be synthesized with modern automated machines if the flanking sequences to which they are to be specifically annealed have already been determined. In the

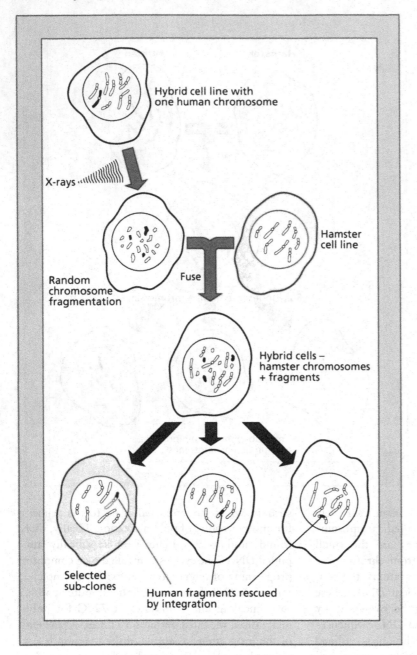

Fig. 5.12 Derivation of cell lines containing fragments of a specific human chromosome. A panel of such lines can be used in the construction of a radiation hybrid map of the chromosome (see p. 153).

case of human DNA, many chromosome-specific sequences can be amplified, even in the absence of prior sequence information, by using *Alu* sequences (see p. 138) for annealing the primers. These short interspersed repeats are distributed so abundantly in the human genome (for example, see Fig. 5.17a) that they are often separated by only a kilobase or less of DNA. By using a primer made to match one end of the *Alu* element, any human chromosome segment flanked by *Alu* elements in the appropriate opposite orientation and no more than a few kilobases apart will be amplified (Fig. 5.13b). If the initial DNA comes from a chromosome-sorted preparation, or from a human–rodent hybrid cell line containing only a single human chromosome or chromosome segment, numerous sequences from that segment can be amplified and cloned to provide probes for the identification of genetic markers or establishment of sequence-tagged sites (see p. 151) for physical mapping of that particular part of the genome.

The PCR technique has become an extremely powerful and versatile tool in molecular genetics. Some of its other uses will be mentioned as we go along.

Comprehensive genetic mapping with molecular markers

Molecular markers present in populations – RFLPs

Linkage maps of chromosomes have traditionally been made by analysis of progenies segregating with respect to gene differences with clear phenotypic effects. The use of such markers has permitted the establishment of linkage groups corresponding to all the chromosomes of such species as *D. melanogaster*, yeast, maize, mouse and humans, to name but a few. However, any attempt to define the complete genome requires a much greater marker density. It turns out that outbreeding species usually have a great reservoir of covert molecular variation, sufficient to supply as many markers as anyone could want if only there were means for detecting it all. Much of this variation is in the form of individually uncommon deviations from a more or less standard type, but a great deal is due to the existence at many loci of several alternatives, all of them fairly common. The latter situation is called *polymorphism*.

Cryptic molecular variation can be detected by analysis either of proteins or of DNA. In the 1960s there was an explosion of interest in enzyme polymorphisms, which were explored most intensively in humans and *Drosophila* species. Where an enzyme can be specifically stained for activity in an electrophoretic gel, the opportunity exists for detecting differently charged forms corresponding to different alleles of the encoding gene. Many enzymes do show this kind of variation in populations (see Chapter 6, p. 171). Enzyme polymorphisms are still interesting and useful in population studies, but for detailed chromosome mapping the emphasis has shifted to variation that can be observed directly in the DNA.

Polymorphism at the DNA level becomes evident when cloned genomic DNA fragments are used to probe Southern blots of restriction digests of DNA samples. Any genomic probe of a few kilobases in length has a good chance of detecting differences in restriction fragment lengths between any two individuals that are not closely related by descent. These differences (*restriction fragment length polymorphisms*; RFLPs) have several origins.

They may be due to single base-pair changes that either create or eliminate restriction sites. Such changes will seldom affect the phenotype if they occur in spaces between genes and, even if they occur within a coding sequence they will not always affect the structure of the protein. For example, a change from AAGCTC to AAGCTT creates a new *Hin*dIII site, but the new sequence, transcribed into mRNA, still codes for lysine—leucine as before.

Alternatively, changes in restriction fragment length may be due to insertions or deletions increasing or decreasing the distance between two restriction sites. Repetitive sequences in tandem array are especially prone to changes in copy number by replication errors or out-of-register crossing over (cf. Fig. 5.6a). The human minisatellite sequence has already been mentioned as a source of hypervariability (p. 132), and much the same could be said of a number of even simpler sequences – tandem repeats of 2- or 3-bp units – that are widely dispersed in the human genome. Probes for sequences with such repeats are always likely to detect differences between individuals.

Probes for RFLPs may recognize either multiple or unique loci. If the restriction fragment variation is due to different-length arrays of minisatellite or other tandem repeat, one can use the repeat itself as a probe. This is likely to detect several different polymorphisms at the same time, which may be an advantage, or possibly a disadvantage if the pattern of hybridized bands becomes too dense to analyse. In the latter event, or if the RFLP arises from a mutational change in a single-copy sequence, it is necessary to use a probe that will hybridize only to one specific chromosome locus.

When the object is to make a detailed map of a specific chromosome, it is best to look for RFLPs with a collection of chromosome-specific probes, which can be made from chromosomes sorted

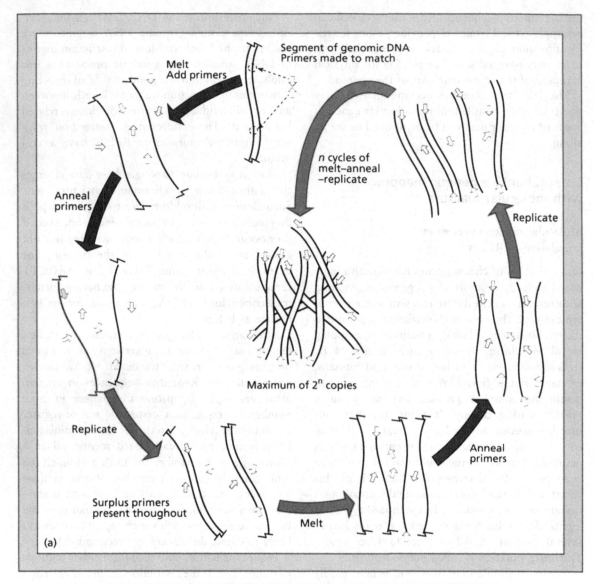

Segment of genomic DNA
Primers made to match

Melt
Add primers

n cycles of
melt–anneal
–replicate

Replicate

Anneal
primers

Maximum of 2^n copies

Replicate

Anneal
primers

Surplus primers
present thoughout

Melt

(a)

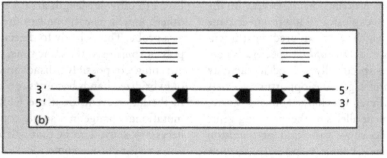

3'
5'
5'
3'

(b)

by one of the methods outlined in the previous section.

Standard genotypes for genetic mapping

Given the large amount of DNA polymorphism in outbred species, any cross-bred diploid should give meiotic segregation of many potential markers on every chromosome. The information available from a segregating family or pedigree is potentially very great, but limited by the current availability of probes. It thus makes sense to preserve recombinant genotypes, or at least standard parental genotypes for generating recombinants; they are likely to provide more information in proportion to the number of new probes that become available.

Haploid analysis in fungi

The general principle can be applied most straightforwardly in haploid fungi, where meiotic products from a cross can be grown and preserved as separate pure cultures. In the important experimental fungus *N. crassa*, probes have been obtained for numerous polymorphic markers, mapped at defined positions within all seven linkage groups, that distinguish two strains of different geographic origins. Several dozen haploid progeny of the cross have been scored with respect to these markers and preserved for reference. Any new *Neurospora* genomic clone can be used to probe the DNA of the parental strains with a good chance of finding a restriction fragment length difference between them. Scoring the set of reference progeny for the new marker will reveal its linkage relationships with the previously established markers and allow it to be assigned to a place on one of the chromosome maps.

Recombinant inbreds

In self-fertile diploid plants, a different procedure can be used to the same end. The haploid products of meiosis cannot be preserved as such, but a feasible alternative is to use individual F_1 plants from a cross between unrelated parents to found a series of recombinant inbred (RI) lines by successive generations of self-pollination. The principle is explained in Fig. 5.14. After several generations of selfing the genotypes are largely homozygous, and each RI line can be regarded for practical purposes as carrying a single recombinant genome and can be scored as if it were a haploid. All markers distinguishing the two original parental strains, whether molecular or directly visible in the phenotype, can be scored without any masking by dominance. A collection of say 50 RI lines from the same initial cross will provide information for mapping of all the loci in which the parent strains can be shown to differ. Note that the primary frequency of meiotic recombination (r) between two loci is less, by a factor of up to two, than the fraction (R) of RI lines in which they are recombined; the exact formula is $r = R/2(1 - R)$. The reason for the difference is that, with multiple generations of self-fertilization, there are repeated opportunities for recombination before all the markers become homozygous.

Through the application of the RI method, between 20 and 50 DNA markers have been mapped on each of the 10 chromosomes of maize (Burr & Burr, 1991).

Using a standard back-cross

Recombinant inbreds are not so easily obtained in experimental animals because of the impossibility of self-fertilization. However, genetic differences

---

Fig. 5.13 (*opposite*) (a) The principle of the polymerase chain reaction (PCR). The reaction mixture contains genomic DNA, deoxynucleoside triphosphate DNA precursors, heat-resistant DNA polymerase from a thermophilic bacterium (usually *Thermophilus aquaticus*, Taq), and an excess of synthetic oligonucleotide primers (short thick arrows) made to match sequences flanking the genomic segment it is desired to amplify. The mixture is subjected to a cycle of temperature changes: for example 1 minute at 91 °C for melting to single strand, $\frac{1}{2}$ minute at 58 °C for annealing of primers and 2 minutes at 72 °C for primed DNA synthesis. Ideally the targeted DNA segment should double in quantity with each cycle. After 25 or 30 cycles it is usually by far the most abundant DNA sequence present. (b) Use of PCR to amplify segments of DNA from a specific sorted human chromosome, for example in a human–rodent hybrid cell line, using primers (small arrows) matching the abundant *Alu* elements (filled boxes) found specifically in human DNA. Any segment of up to 1–2 kb in length, and flanked by *Alu* elements in the appropriate inverted orientation, should be amplified. After Bicknell *et al.* (1991).

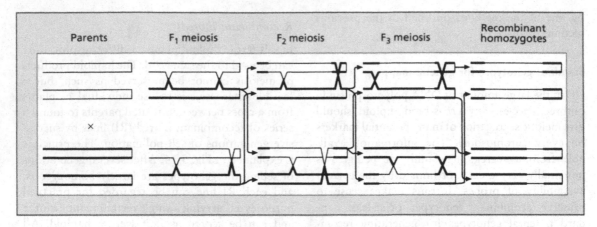

Fig. 5.14 Derivation of recombinant inbred lines in a self-fertilizable plant such as *Zea mays*. A cross is made between two distantly-related inbreds, differing in many molecular markers (chromosomes distinguished by thick versus thin lines) and F_1 plants are used to found a number of lines each maintained by self-pollination of a single plant in each generation. The diagram illustrates the progress towards homozygosity of just one chromosome. The boxes enclose hypothetical patterns of crossing over, in an ovule and an anther of a single plant in each case, that precede the formation of the egg and pollen nuclei that unite to produce the single plant used in the following generation. After a number of generations, each recombinant inbred will be homozygous for its own particular combination of segments from the two original lines. The particular crossover positions shown are chosen arbitrarily for purposes of illustration. Burr & Burr (1991) report the mapping of 334 molecular markers spread over all 10 chromosomes, based on the frequencies of separation of originally linked markers in approximately 100 recombinant inbreds.

of potential use in mapping can be kept in the form of distantly related inbred lines. In one analysis in mouse, a laboratory strain of *Mus musculis* was crossed to a strain of *M. spretus*; the two were fully inter-fertile but differed with respect to many RFLPs. F_1 mice were back-crossed to the *M. musculis* strain and DNA samples from 114 backcross progeny were screened with a set of probes specific for one segment of chromosome 2, and scored for presence or absence of the *M. spretus* markers. As explained in Fig. 5.15, 14 markers were placed in order on chromosome 2 through this single back-cross.

As this example shows, DNA markers have the advantage over most phenotypic differences not only in their greater availability but also because their scoring is not subject to the problem of dominance. Alleles recognized at the DNA level are codominant – that is independently recognizable in heterozygotes. A recessive allele from one species could not have been scored in the progeny of the back-cross to the other species.

Preserved pedigrees

In human genetic mapping, geneticists have to take pedigree data as they find it. Controlled crosses are ruled out, but inferences can be drawn so long as relationships are known. Since any family is certain to show segregation of a great many molecular markers if only probes can be found for them, it is a reasonable strategy to store cell cultures from members of human pedigrees and score them for additional markers as the opportunity and need arises. One such reference pedigree collection has been established in France at the Centre d'Étude Polymorphisme Humaine (CEPH). It consists of cells from 40 different families from different countries, selected for large size – the average number of sibs per family is about eight.

For mapping of one chromosome or chromosome segment, a set of probes specific for that segment is selected and used to screen parental DNA samples to find out which families should be segregating with respect to two or more of the polymorphisms revealed by the probes. Similar

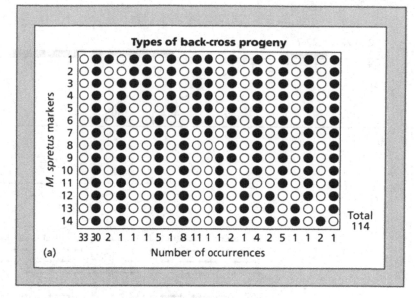

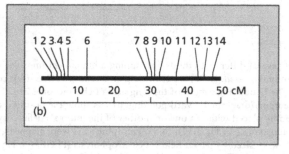

Fig. 5.15 (a) Genetic mapping of molecular markers in mouse by means of an inter-specific cross and back-cross. The laboratory strain of *Mus musculis* was found to differ from the *M. spretus* strain in 14 independently scorable DNA markers on chromosome 2. In total, 114 progeny of the backcross of the hybrid to *M. musculis* were scored for presence (●) or absence (○) of each of the *M. spretus* markers.

(b) The data were used to deduce the map, with distances between successive markers in centimorgans (percentage recombination). Note that the complete inter-fertility shown by *M. musculis* and *M. spretus* is not usual for forms placed in different species. Data from Rochelle *et al.* (1992).

probing of DNA samples from the children of such families provides data that can be evaluated by Lod analysis (p. 29). This complicated task is delegated to a computer programme which calculates the most probable sequence of the linked markers and the best estimates, in terms of centimorgans, of the distances between them.

This kind of analysis is greatly facilitated by amplification of the polymorphic sequences by PCR, using appropriate probe sequences as primers. In this way the RFLPs can be scored starting with minimal amounts of DNA.

The same pedigrees, preserved in the form of cell cultures, are being used for mapping all the human chromosomes (see the Report of the NIH/CEPH group, 1992). Some chromosome segments have been mapped with continuous series of markers no more than 2 cM apart. The small sample of chromosome 21 markers shown in Fig. 5.16 were selected because they were also

positioned on physical maps (see below).

A point of interest arising from such detailed human genetic mapping is that recombination frequency in human female meiosis is about 40% higher than in the male. Thus the conventional unit of 1% recombination, the centimorgan, does not have the same meaning in the two sexes. Standard human genetic maps tend to show sex-average distances.

Physical mapping

If a chromosome can be resolved physically into fragments, it is possible to map it without the help of natural recombination or genetic variation. All that is necessary is a large collection of probes for recognizing specific chromosome sites – *sequence-tagged sites* or STSs. Chromosome fragments sharing the same tagged sites clearly overlap, and on this basis the fragments can be placed in order.

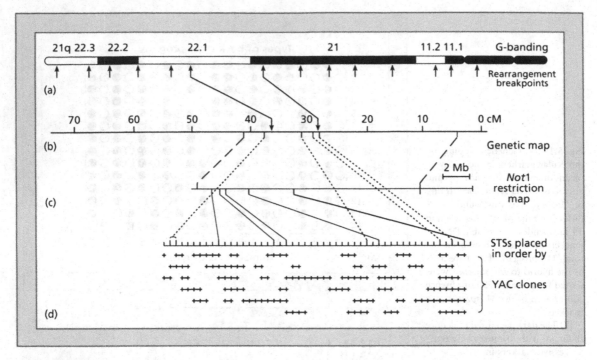

Fig. 5.16 Correlation of several different methods of mapping a human chromosome using sequence-tagged sites (STSs): a sample of the data of Chumiakov *et al.* (1992) concentrating on a central 25% of their complete physical map of the long arm of chromosome 21. (a) The G-banding pattern of the long arm of chromosome 21 (21q) and the numbered subdivisions so defined are shown in the upper diagram. Upward-pointing arrows show the positions of the breaks in a series of segmental rearrangements. STSs can be placed within in one or another of the intervals defined by the breakpoints by hybridization to isolated rearranged chromosomes (see Fig. 5.11). (b) Polymorphic STSs were mapped genetically by Lod analysis of pedigree data. (c) *Not*1 fragments were sized by pulsed-field gel electrophoresis (PFGE) (cf. Fig. 5.9) and ordered by means of *Not*1 linking clones (see p. 153). Some STSs were placed near *Not*1 sites by their hybridization to linking clones. (d) Several hundred chromosome 21-specific yeast artificial chromosome (YAC) clones were screened for presence or absence of each of 198 STSs, some of which had been mapped by one or more of the other methods. YAC clones were placed in a contiguous series ('contig') by the overlaps between their arrays of STSs. A part of the YAC contig representing the greater part of 21q22.1, overlapping into 21q21, is shown. Each YAC clone is shown as a line with a series of arbitarily evenly spaced dots, each dot being an STS. Only a sample of the YACs carrying two or more STSs is shown to illustrate the way they were ordered by their overlaps. The average size of the YAC clones is about 1 megabase, the average spacing of STSs about 200 kb, and the total length of 21q is estimated as 40–50 megabases. The analysis was also assisted by the location of some STSs on a radiation hybrid map (see p. 153 and Fig. 5.12) – not illustrated here.

The following account is largely based on methods that have been developed in connection with the human genome project, though most of them are applicable more generally. Different levels of resolution are obtained with different degrees of chromosome fragmentation, ranging from large sections of chromosome arms down to fragments of DNA small enough to be propagated in cloning vectors. Some results of a combination of methods applied to human chromosome 21 are shown in Fig. 5.16.

Locating DNA sequences on the microscopically observable chromosome

Many different segmental interchanges have been recorded in the human genome, and a number of the rearranged chromosomes have been preserved in hybrid cell lines. Each chromosome is then effectively divided into a number of segments delimited by its rearrangement breakpoints. Probing to determine which rearrangements include a particular STS and which do not define the segment within which the STS is located. This enables STSs to be positioned in relation to the chromosome banding pattern – the cytological map (Fig. 5.16a). Another means to the same end is fluorescent labelling of gene loci (the FISH technique, see p. 91).

Cytological mapping of genes is (except in *Drosophila*) a very coarse level of analysis, but at least it establishes some clearly ordered landmarks, the spaces between which can later be filled in with further tagged sites.

Not1 fragments and linking clones

Restriction enzymes, such as *Not*1, which recognize 8-bp palindromes, cut DNA into fragments ranging from tens of kilobases to several megabases. The fragments derived from a particular human chromosome (isolated in a hybrid cell and recognized by a probe for the ubiquitous *Alu* sequence) can be separated and sized by pulsed-field gel electrophoresis. To build them into a continuous map it is necessary to find smaller fragments to bridge between their ends. Such linking fragments can be identified, for example, in a *Hin*dIII library cloned in a lambda vector – they will be that small minority of *Hin*dIII fragments with internal *Not*1 sites. Each linking clone will hybridize with two *Not*1 fragments, so demonstrating that they are adjacent to each other.

The linking clones also enable some STSs to be placed on the *Not*1 map, since any STS present in a short linking clone must be close to the corresponding *Not*1 site. As shown in Fig. 5.16, the STSs located in this way can also be mapped in relation to the chromosome banding pattern and, ultimately, in a much finer-scale map based on YAC clones – see below. The usefulness of a *Not*1 map lies in the fact that it is calibrated in terms of real distance and so provides a physical scale for the other kinds of map.

Radiation hybrid mapping

The principle of using radiation hybrids for separating human chromosome fragments was explained above (see Fig. 5.12). Given a large number of hamster cell lines each containing a fragment of a particular human chromosome, it is possible to use PCR to amplify known STSs and so determine which STSs are present on which fragment. Distances between sites can then be expressed in units ('centirays') representing the percentage probability of separation by breakage with a given X-ray dose. In fact, this gives a better measure of physical distance than meiotic recombination frequency, expressed in centimorgans. Whereas recombination frequency per unit of DNA varies markedly between different chromosome regions, the vulnerability to X-ray breakage seems to be fairly constant along the whole length of the chromosome.

Contiguous series of YAC clones

The use of yeast artificial chromosomes (YACs) for the cloning of chromosomal DNA fragments of up to a megabase (1000 kb) or more in length was outlined in Chapter 3. Vectors with smaller carrying capacities, such as cosmids, are still useful for the analysis of relatively small genomes but YACs are now the strongly favoured means for the continuous end-to-end physical mapping of human chromosomes.

The general method is to start with a large set of YAC-cloned fragments, preferably from a single chromosome sorted by one of the methods outlined above. Overlaps between different cloned fragments are then established on the basis of shared STSs. In the work leading to the map summarized in Fig. 5.16, the entire long arm of human chromosome 21 was cloned in 180 YACs including about 200 STSs. YAC clones were screened for STSs by the PCR procedure, using appropriate primers for specific amplification of each STS. The results enabled all the YAC clones, and hence the STSs, to be placed in a contiguous series ('contig') on the basis of their overlapping sets of tagged sites. Figure 5.16 shows a portion of the contig, comprising just 50 STSs, to illustrate the principle.

From physical map to complete DNA sequence

Different gene densities in different organisms

Once the genome, or a section of it, has been cloned in manageable ordered pieces, it becomes possible to obtain the complete DNA sequence. Whether this is done or not is a matter of priorities and finance.

Various total genome projects are at present under way. The farthest advanced is that on *Escherichia coli*, which is still, at the time of writing, only about 40% complete. In the eukaryotes, one of the 18 yeast chromosomes, chromosome 3, has been completely sequenced. Physical genomic mapping preparatory to DNA sequencing is being most intensively pursued for humans (Olson, 1993), but is also well advanced in *Drosophila* (Hartl *et al.*, 1992), the nematode *Caenorhabditis elegans* (Sulston, 1992) and the plant *Arabidopsis thaliana*. Sundry chromosome segments of the order of 100 kb in length have been completely sequenced in a variety of species. Different organisms show striking differences in their patterns of gene organization (Fig. 5.17).

The *E. coli* genome, a representative piece of which is shown in Fig. 5.17e, is packed with genes at minimal spacing. There are no introns, and different open reading frames are often included in the same transcription unit, or operon.

In budding yeast (Fig. 5.17c), there tends to be more space between the open reading frames, but the organization is still relatively compact. Introns are absent from most genes and, where they occur, are seldom more than 100 bases in length. Interspersed among the genes are occasional transposable retro-elements of the Ty family and, more numerous isolated copies of their long terminal repeats (δ elements), presumably relics of complete Tys that have excised by recombination (see Fig. 5.4b).

So far as it has been sequenced, the *Caenorhabditis* genome (Fig. 5.17d) has a gene density not very different from that of yeast, and introns, though numerous, are still fairly short.

Mammalian genomes, as exemplified by mouse and human, are greatly expanded as compared with the small genomes just considered. There are more genes, and they are, on the whole, much more thinly spread. Exons are often short and separated by relatively enormous introns, often kilobases or even tens of kilobases long. The introns, as well as the spaces between the genes, contain great numbers of dispersed repeats, SINES and LINES. The most abundant human SINE, the *Alu* element, is particularly prominent in the hypoxanthine phosphoribosyl transferase (HPRT) locus, shown in Fig. 5.17a, but other parts of the human genome have LINEs at a density comparable to that of the mouse L1 element in the β-globin gene cluster in mouse (Fig. 5.17b). In other regions, especially those associated with heterochromatin, there are few genes and great blocks of simple repetitive ('satellite') sequence.

Fig. 5.17 (*opposite*) Some contrasting chromosomal landscapes. In each example, gene coding sequences, either of confirmed function or (in the yeast example) identified as open reading frames (ORFs), are shown as black bars or boxes in the upper line, with transcripts and their orientation shown by the arrows above; interspersed repetitive elements are shown separately below. The proportion of the total genome represented by the fragment illustrated is shown to the right of each diagram. (a) A 60-kb region of human chromosome 5 containing the *HPRT* (hypoxanthine phosphoribosyl transferase) gene. The protein-encoding sequences are widely dispersed in nine relatively short exons. The long introns and the flanking sequence contain abundant short dispersed repetitive sequences, mostly *Alu* elements, but some (L) highly shortened versions of the most abundant human long interspersed element (LINE). After Edwards *et al.* (1990). (b) A 60-kb region of mouse chromosome containing the β-globin gene cluster. The introns are of moderate length. The spaces between genes consist to the extent of about 30% of LINE (L1) elements truncated to various degrees (some are complex patchworks of L1 sequences). After Jahn *et al.* (1980) and Hutchison *et al.* (1989). (Note that the differences in intron size and kinds of repetitive sequence seen between A and B could be found within either species, and are not meant to imply an overall difference between man and mouse.) (c) A 60-kb section of the complete 315 kb DNA sequence of *Saccharomyces cerevisiae* chromosome 3. Genes and candidate genes (*LEU2*, *PGK1* and numerous unidentified ORFs) are closely packed and contain few if any introns; directions of transcription are indicated by arrows. There are some transfer RNA genes (Pro, Asn, Glu), occasional Ty retroelements and more numerous isolated Ty long terminal repeats (δ). After Oliver *et al.* (1992). (d) A 40-kb *Caenorhabditis elegans* sequence cloned in a cosmid. Protein-encoding sequences are about as closely packed as in *Saccharomyces*, but with numerous short introns. This sequence includes one transposon (TcA) – there are of the order of 100 in the whole genome. After Sulston *et al.* (1992). (e) A 50-kb section of the *Escherichia coli* genome. ORFs are packed virtually end-to-end, and several are often included in the same unit of transcription (operon). A little over half of them are known genes. The origin of replication (*ori*) is indicated. After Burland *et al.* (1993).

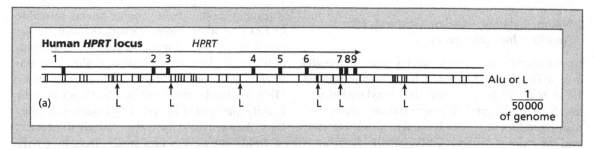

Human *HPRT* locus

HPRT

1   2 3   4   5   6   7 8 9

Alu or L

(a)   L   L   L   L L   L

$\dfrac{1}{50\,000}$ of genome

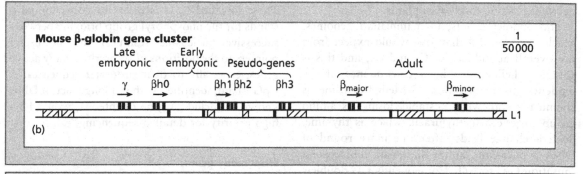

Mouse β-globin gene cluster

$\dfrac{1}{50\,000}$

Late embryonic   Early embryonic   Pseudo-genes   Adult

γ   βh0   βh1 βh2   βh3 β_{major} β_{minor}

(b)   L1

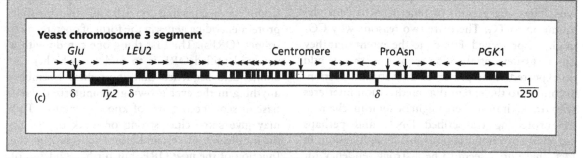

Yeast chromosome 3 segment

Glu   *LEU2*   Centromere   ProAsn   *PGK*1

(c)   δ   *Ty2*   δ   δ

$\dfrac{1}{250}$

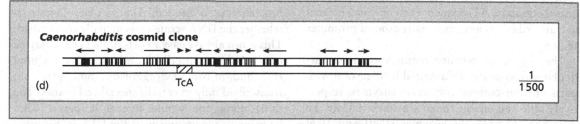

***Caenorhabditis* cosmid clone**

(d)   TcA

$\dfrac{1}{1500}$

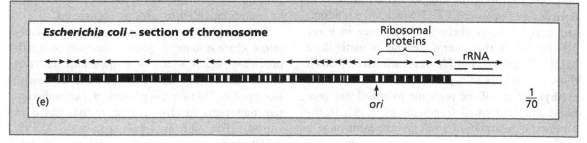

***Escherichia coli* – section of chromosome**

Ribosomal proteins

rRNA

(e)   *ori*

$\dfrac{1}{70}$

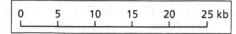

0   5   10   15   20   25 kb

DNA methylation – undermethylated 'islands' as markers for active genes

In mammals and many flowering plants, though not in *D. melanogaster* and fungi, chromosomal DNA tends to be quite heavily methylated along much of its length. The methylation affects cytosine residues in 5'-CG-3' sequences in mammals and in both CG and CXG (where X is any deoxynucleotide residue) in plants.

The doublet CG is, in mammalian genomes, much less plentiful than one would expect from the overall abundance of Cs and Gs, and this is generally believed to be a consequence of the tendency to methylation. 5-Methyl-cytosine is chemically somewhat unstable, losing its amino group to give 5-methyluracil, which is thymine. Such a change leads, after one more round of DNA replication, to the base-pair replacement mutation G/C to A/T. Consequently CG doublets will, over long periods of time, tend to decay by mutation to TG. There are two reasons why CGs might be preserved. Firstly, to the extent that they are part of essential codon sequences, they would be expected to be preserved by natural selection. Secondly, to the extent that methylation interferes with transcription, there might be some mechanism for protecting transcribed DNA, and perhaps especially promoter regions, from methylation. In fact, there does seem to be a strong tendency for non-methylated CG sequences to be concentrated in transcribed regions, especially around promoter sequences.

The degree of cytosine methylation in a particular sequence can be assessed by digestion with pairs of 'four-cutting' restriction enzymes, respectively able and unable to cut a methylated site. To take one example, the enzymes *Hpa*II and *Msp*I both cut the sequence CCGG, which, on a random-sequence basis, is likely to occur once in every 50–100 bp. If the inner C residue is methylated *Msp*I will still cut, but *Hpa*II is ineffective. Thus if the CG doublets in a particular DNA region are methylated it will be resistant to *Hpa*II but probably cut into small fragments by *Msp*I; in the absence of methylation it will give the same pattern of fragments with both enzymes. The difference can be seen very easily by appropriate probing of Southern blots of paired *Msp*I and *Hpa*II digests.

As probes for the whole genome become available in YACs or other cloning vectors, the methylation status of any part of it can be readily determined.

When this kind of analysis is applied to mammalian genomes it is found that long tracts of DNA are hardly cut at all by *Hpa*II and must be heavily methylated in such CG sequences as they have. Between these there are relatively short tracts of DNA that are cut into small pieces by *Hpa*II. These so-called *CpG islands* (the conventional p stands for the phosphoryl group bridging between successive nucleoside residues) can be useful pointers to the positions of transcriptionally active genes, especially to their promoter sequences. A CpG island, identified within a long tract of DNA known to include a gene of interest, would be a high priority for detailed sequencing.

How many genes?

Complete sequencing reveals all the possible protein-encoding genes in the form of open reading frames (ORFs). The first thing one can do with a newly-identified ORF is to find out whether the amino acid sequence that it predicts resembles anything in the ever-growing computerized database of sequenced genes of known function. This may give some clue, strong or weak depending on the degree of resemblance discovered, to the function of the new ORF, but further evidence of functionality is needed. The second question is whether the DNA sequence is actually transcribed. This is not always easy to determine; genes may be only weakly transcribed and still play essential roles and, in complex organisms, many genes are transcribed only in certain specialized tissues. The third thing one would like to know is what happens to the organisms if the ORF is disrupted by mutation, either natural or engineered.

The complete sequence of *Saccharomyces cerevisiae* chromosome 3 goes some way towards providing answers to these questions for this unicellular eukaryote, which has the advantage (see pp. 88–90) that it is possible to target disruptive mutations to any yeast gene that has been cloned. The chromosome sequence (315 kb, 2.25% of the total genome) was found to contain 182 ORFs for polypeptides longer than 100 amino acids, of which 37 corresponded to already known yeast genes and 29 others showed some resemblance to known genes from other organisms. There was no comprehensive and detailed information as to how many of the ORFs were transcribed, but, from other estimates of the complexity of the whole population of yeast transcripts, it seemed not unreasonable to suppose that most if not all of them were. The surprise came when a sample of 55 of the unknown ORFs were individually disrupted. The disruption was lethal to the haploid cell in only three cases. Of the other disruptions, 42 were examined carefully and 14 found to be associated with slow growth or other non-lethal effect; the remaining 28 had no discernible effect at all. This is rather puzzling, since if all these ORFs were really functionless like pseudogenes, one would not have expected them to have been preserved.

If we take 100 as an approximate estimate of the number of genes with necessary functions on yeast chromosome 3, and if that chromosome is representative of the whole genome, we can conclude that *Saccharomyces* has about 4000 to 5000 essential genes. But there would seem to be nearly as many again with duplicative or 'back-up' functions, or functions needed only under special conditions, that would not be easily uncovered in conventional mutant hunts.

In the *Drosophila* genome there are approximately 5000 polytene chromosome bands. Several detailed studies (see p. 50) have suggested that the numerical relationship between polytene chromosome bands and genes capable of mutating to recessive lethality is one-to-one, or at any rate one to no more than two. So there may be 5000–10 000 single-copy genes with indispensable functions. However, we can expect, in *Drosophila*, at least as high a proportion of genes with 'redundant' or 'back-up' functions as in yeast.

There is even less certainty about gene number in humans. Individual human genes, with their introns, seem typically to occupy 50 kb or so, which would make room for 60 000 genes, or fewer if heterochromatin is excluded from the functional genome. A better estimate may be obtainable for the number of transcription-initiation sites, which seem in general to correspond to methylation-free 'islands'. This number appears to be about 100 000.

DNA of the organelles

The closed-loop DNA molecules present in eukaryotic organelles – the *mitochondria* and, in green plants, the *chloroplasts* – are essential genomic components of eukaryotes. Their sequences are far less complex than those of the chromosomes and much more susceptible to complete analysis. They are the easiest part of a genome project.

Mitochondrial genomes – total analysis

Mitochondria, which are the organelles of respiration, are typically present in eukaryotic cells in tens or even hundreds, and each mitochondrion contains several copies of a specific small mitochondrial genome (mtDNA). The first mitochondrial genome to be characterized was that of yeast, *S. cerevisiae*, and it is the only one to have been mapped by the classical genetic methods of mutation, complementation and recombination before the advent of DNA sequence analysis.

The initial analysis of yeast mtDNA was based on an extensive series of respiration-deficient mutants, already mentioned in Chapter 1 (see p. 43). These mutants owe their respiratory deficiencies to more or less extensive deletions of the mtDNA. Diploids formed by crossing haploid strains carrying different mitochondrial deletions are competent in respiration provided that each mutant carries the mitochondrial genes missing in the other (a complementation test). Furthermore, if the two deletions do not overlap, recombination between mtDNA molecules, which occurs with a surprisingly high frequency during the first few mitotic divisions of the diploid hybrid, will yield a certain frequency of recombinant clones with wild-type mitochondria. Mutant mitochondria with overlapping deletions cannot recombine to give wild-type recombinants.

The overlaps defined by these genetic tests enabled the segments deleted in the different mutants to be arranged in a circular map. This result accorded with restriction-site mapping of the mtDNA, which also demonstrated a closed loop of about 75 kb. Gene functions could be assigned to different parts of the restriction map by correlating retained restriction fragments with retained

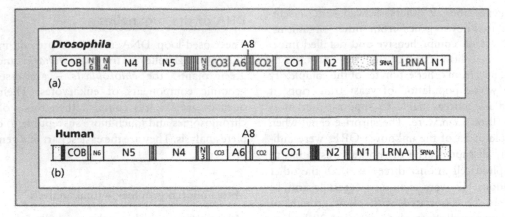

Fig. 5.18 Mitochondrial genomes. (a) *Drosophila subobscura*. (b) Human. The circular DNA molecules are shown as linear for convenience in drawing; the ends should be imagined as joined. The gene products are as follows: COB, cytochrome b; CO1, 2, 3, subunits of cytochrome c oxidase; A6, 8, subunits of ATPase; N1, 2, 3, 4, 5, 6, subunits of NADH dehydrogenase; SRNA, LRNA, small and large ribosomal RNAs; narrow stripes are transfer RNA genes; stippled boxes are the only non-coding regions and contain origins of replication. The total DNA is about 16 and 16.5 kb for fly and human, respectively. (a) After Volz-Lingenhohl *et al.* (1992) and (b) after Anderson *et al.* (1981).

genetic functions, as determined by complementation tests with point mutants with defined deficiencies.

mtDNAs have now been completely sequenced from several different species, including both plants and animals, and all their genes have been identified. Although differing in detailed architecture, they are functionally largely similar. They include only a very limited set of genes, encoding some, but by no means all, of the proteins and RNA molecules found in the mitochondria themselves. Mitochondria have their own special translation machinery, with rRNAs and a complete set of tRNAs all distinct from the corresponding components of the cytoplasmic apparatus. The genes for mitochondrial ribosomal proteins are chromosomal, but the rRNAs and tRNAs are transcribed from the mtDNA. Other mtDNA genes encode components of respiratory pigments and enzymes – cytochrome b, and a number of the subunits of the complex enzymes reduced nicotinamide adenine dinucleotide (NADH) dehydrogenase, cytochrome c oxidase and ATPase, all involved in the reoxidation of reduced respiratory intermediates coupled to the generation of chemical energy in the form of adenosine triphosphate (ATP).

Animal mtDNAs are organized in an extremely compact fashion, with almost no spaces between the coding sequences. Nearly all the boundaries between genes are marked by tRNA genes, a fact which is significant in connection with RNA processing. The mtDNA yields a single RNA transcript, from which the tRNA molecules are excised. The fragments left are predominantly messengers for single protein-coding genes. Figure 5.18 shows a comparison between human and *Drosophila* mtDNAs; the high degree of similarity between the two, in size, gene content and even, to a large extent, in gene order is somewhat astonishing.

The mtDNAs of fungi, now known from several species, tend to be much longer than those of animals – 75 kb in budding yeast, 63 kb in *Neurospora* and 94 kb in another filamentous ascomycete, *Podospora anserina*. They contain very nearly the same set of genes as animal mtDNAs, or perhaps fewer in yeast, where the NADH dehydrogenase genes have not been identified as yet. The extra length of fungal mtDNA is accounted for partly by the presence of more or less abundant self-splicing mitochondrial introns (see p. 159) and also by non-coding spacer DNA between the genes. Most of the spacer sequence in the yeast mitochondrial genome consists of A/T-rich repetitive sequences with shorter G/C-rich repeats.

The mtDNAs of green plants, so far as they have been analysed, are different again. In maize (*Zea mays*), for example, the mtDNA consists of upwards of 500 kb of unique sequence, not all necessarily in the same molecule but sometimes, at least, distributed between a number of different DNA circles. It appears to contain more or less the same complement of genes as animal mtDNAs but with a great deal of redundant or spacer sequence.

The mitochondrion as an alien system

The mitochondrial genetic system is in several ways very different from that of the cell nucleus. In their closed-loop DNA structure, lack of nucleosomes, the sizes of the rRNAs and the susceptibility of the ribosomes to antibacterial antibiotics, mitochondria seem to have bacterial affinities. It is speculated that they may have originated from bacteria that formed a symbiotic relationship with the primaeval eukaryotic cell, but they have some peculiarities not shared by modern bacteria. Whereas eukaryotic nuclear genes and typical bacteria share the 'universal' genetic code (Table 2.2), mitochondrial genes show several deviations. In vertebrate animal mitochondria for example, ATA is a methionine codon and AGA and AGG are not used as codons at all. UGA, which is a terminator in the standard code, is a tryptophan codon in mitochondrial genes of all eukaryotes so far as they have been sequenced.

Mitochondrial genes also tend, at least in fungi, to have their own special kinds of introns. These differ from the introns of chromosomal genes in several ways, the most interesting of which is that they catalyse their own excision. The interaction between different parts of the intron sequence, on which the chemistry of excision depends, is often stabilized by a protein encoded in the intron itself.

The degree of autonomy enjoyed by the mtDNA can, on occasion, be an embarrassment to the organism. In both yeast and in various filamentous fungi, mutant mtDNA molecules that have lost essential functions by deletion often turn out to be more efficient in replication than their wild-type progenitors, which they tend to displace as growth proceeds. In *Saccharomyces*, deletions in mtDNA appear to be the result of recombination between repeats in the spacer regions. In various filamentous fungi, the senescence that sometimes overtakes vegetatively propagated strains is usually caused by efficiently replicating defective mtDNA which arises, as in yeast, by internal recombination. In maize, 'rogue' variants of mtDNA have been shown to cause pollen sterility.

Chloroplast DNA

In green leaves, DNA contained in the chloroplasts, the organelles of photosynthesis, exceeds in mass that of the cell nuclei. The cells are packed with chloroplasts and each chloroplast contains numerous copies of a relatively small closed-loop molecule — 120–160 kb in the plants analysed so far.

Formal genetic analysis of the structure and functions of chloroplast DNA has been attempted seriously only in the unicellular green alga *Chlamydomonas reinhardi* and we shall not be dealing with that interesting but somewhat esoteric genetic system in this book. So far as flowering plants are concerned, everything that we know in detail comes from restriction-site mapping followed by DNA sequencing and identification of open reading frames, mainly on the basis of homology with the sequences of RNA and proteins whose functions have been established by other means. What emerges is somewhat analogous to the situation with regard to mitochondria. There are genes for the special apparatus of chloroplast protein synthesis — chloroplast-specific rRNAs and tRNAs and (unlike the mitochondrial situation) for chloroplast ribosomal proteins. A number of the components of the photosynthetic apparatus are also encoded in the chloroplast genome, including the larger of the two polypeptide subunits of the enzyme immediately responsible for carbon dioxide fixation — ribulose biphosphate carboxylase. The smaller subunit is the responsibility of a chromosomal gene.

Conclusion: total analysis of genomes – what benefits to understanding?

Knowing the total sequence of an organism's DNA will be a great help in hunting for specific genes. It will also provide a catalogue of all the polypeptide chains that the organism has the information to produce, but only a provisional catalogue, since alternative modes of splicing-out of introns can yield different polypeptides from the same primary gene transcripts. A complete cDNA library is really needed to supplement the information from the genomic sequence.

A complete amino acid sequence catalogue would, in itself, give only limited and uncertain information about the functions of the proteins formed from all these polypeptide chains. It will not show how the transcription of the genes encoding them is spatially and temporally regulated during development, still less how the thousands of genes and gene products interact with each other.

Helpful as it will be in many ways, the complete sequence of the genome will certainly not be the definitive 'Book of Life' that some imagine. It can hardly be more than the introductory section of something vastly longer and more complicated. Whether that extended treatise can ever be completed, and whether we can at least begin to see what its outlines might be, are questions that will be addressed in the final chapters of this book.

References

Anderson, S., Bankier, A.T., Barrell, B.G. *et al.* (1981) Sequence and organization of the human mitochondrial genome. *Nature*, **290**, 457–65.

Bicknell, D.C., Markie, D., Spurr, N.K. & Bodmer, W.F. (1991) The human chromosome content in human × rodent somatic cell hybrids analyzed by a screening technique using Alu PCR. *Genomics*, **10**, 186–92.

Bingham, P.M. & Zachar, Z. (1989) Retrotransposons and the FB transposon from *Drosophila melanogaster*. In Berg, D.E. & Howe, M.M. (eds) *Mobile DNA*, pp. 485–502. American Society for Microbiology, Washington, DC.

Boeke, J.D. (1989) Transposable elements in *Saccharomyces cerevisiae*. In Berg, D.E. & Howe, M.M. (eds) *Mobile DNA*, pp. 335–74. American Society for Microbiology, Washington, DC.

Britton, R.J. & Kohne, D.E. (1968) Repeated sequences in DNA. *Science*, **161**, 529–31.

Brody, H. & Carbon, J. (1989) Electrophoretic karyotype of *Aspergillus nidulans*. *Proc Natl Acad Sci USA*, **86**, 6260–3.

Burland, V., Plunkett, G. III, Daniels, D.L. & Blattner, F.R. (1993) DNA sequence and analysis of 136 kilobases of the *Escherichia coli* genome: organizational symmetry around the origin of replication. *Genomics*, **16**, 551–61.

Burr, B. & Burr, F.A. (1991) Recombinant inbreds for molecular mapping in maize: theoretical and practical considerations. *Trends Genet*, **7**, 55–60.

Chumiakov, I., Rigault, P., Guillou, S. *et al.* (1992) Continuum of overlapping clones spanning the entire human chromosome 21q. *Nature*, **359**, 380–7.

Craig, J.M. & Bickmore, W.A. (1993) Chromosome bands. *BioEssays*, **15**, 349–54.

Deininger, P.L. (1989) SINES: short interspersed repeated DNA elements in higher eukaryotes. In Berg, D.E. & Howe, M.M. (eds) *Mobile DNA*, pp. 619–36. American Society for Microbiology, Washington, DC.

Edwards, A., Voss, A., Rice, P. *et al.* (1990) Automated DNA sequencing of the human HPRT locus. *Genomics*, **6**, 593–608.

Engels, W.R. (1989) P elements in *Drosophila*. In Berg, D.E. & Howe, M.M. (eds) *Mobile DNA*, pp. 437–84. American Society for Microbiology, Washington, DC.

Fedoroff, N.V. (1988) Maize transposable elements. In Berg, D.E. & Howe, M.M. (eds) *Mobile DNA*, pp. 375–412. American Society for Microbiology, Washington, DC.

Hartl, D., Ajioka, J.W., Cai, H. *et al.* (1992) Towards a *Drosophila* genome map. *Trends Genet*, **8**, 70–5.

Hutchison, C.A. III, Hardies, S.C., Loeb, D.D., Shehee, W.R. & Edgell, M.H. (1989) LINEs and related retrotransposons: long interspersed repeats in the eukaryotic genome. In Berg, D.E. & Howe, M.M. (eds) *Mobile DNA*, pp. 593–618. American Society for Microbiology, Washington, DC.

Jahn, C.L., Hutchison, C.A. III, Phillips, S.J. *et al.* (1980) DNA sequence organization of the β-globin complex in the BALB/c mouse. *Cell*, **21**, 159–68.

Langlois, R.G., Yu, L.-C., Gray, J.W. & Carrano, A.V. (1982) Quantitative karyotyping of human chromosomes by dual beam flow cytometry. *Proc Natl Acad Sci USA*, **79**, 7876–80.

NIH/CEPH collaborative mapping group (1992) A comprehensive genetic linkage map of the human genome. *Science*, **258**, 67–102.

Oliver, S.G., van der Aart, Q.J.M., Agastoni-Carbone, M.L. *et al.* (1992) The complete DNA sequence of yeast chromosome III. *Nature*, **357**, 38–46.

Olson, M.V. (1993) The human genome project. (Review). *Proc Natl Acad Sci USA*, **90**, 4338–44.

Ried, T., Baldini, A., Rand, T.C. & Ward, D.C. (1992) Simultaneous visualization of several different DNA probes by *in situ* hybridization using combinatorial fluorescence and digital imaging microscopy. *Proc Natl Acad Sci USA*, **89**, 1388–92.

Rochelle, J.M., Watson, M.L., Oakey, R.J. & Seldon, M.F. (1992) A linkage map of mouse chromosome 19: definition of comparative mapping relationships with human chromosomes 10 and 11 including the *MEN1* locus. *Genomics*, **14**, 26–31.

Sulston, J., Du, Z., Thomas, K. *et al.* (1992) The *C. elegans* sequencing project: a beginning. *Nature*, **356**, 37–41.

Varmus, H. & Brown, P. (1989) Retroviruses. In Berg, D.E. & Howe, M.M. (eds) *Mobile DNA*, pp. 53–108. American Society for Microbiology, Washington, DC.

Volz-Lingenhohl, A., Solignac, M. & Sperlich, D. (1992) Stable heteroplasmy for a large deletion in the coding region of *Drosophila subobscura* mitochondrial DNA. *Proc Natl Acad Sci USA*, **89**, 11528–32.

6

ACCOUNTING FOR
HERITABLE VARIATION

We are now no longer satisfied by cataloguing and mapping the gene(s) responsible for a particular kind of variation; we would also like, where possible, to determine their sequences and the molecular nature of the mutations that differentiate alleles.

The older ways of detecting and measuring genetic variation are being reinforced and extended by the new. How complete an account of heritable variation is now in prospect?

Kinds of genetic variation

It is convenient to consider genetic variation under three headings.

Firstly, there are, in all outbreeding species, a considerable number of allelic differences the individual effects of which can be easily seen in the phenotype. Some of these, in diploid organisms, are rare deleterious recessives, originating by mutation and persisting in the population because they are protected from counter-selection in heterozygotes. Numerous hereditary diseases in humans come into this category, and all outbreeding diploid organisms have equivalent conditions. Other variations qualify as polymorphisms, meaning that two or more alleles persist in populations at noticeable frequencies, either because they are more or less inconsequential for fitness or because they are maintained by natural selection. In the latter case, the mechanism may be frequency-dependent selection, that is selection in favour of the less frequent genotypes, as is suspected to occur with human blood groups and other components of the immune system. Or, probably more rarely, the advantage may lie

Introduction

Background

The science of genetics began with the aim of explaining natural heritable variation. In recent times, this initial objective has been somewhat overshadowed by the increasing emphasis on the analysis of standard ('wild type') gene structure and function, which has become partly detached from the study of variation. Nevertheless, inherited variation remains central to genetics. It is the key to the short-term ecological adaptation and long-term evolution of all forms of life. Naturally occurring variation still, in spite of the growing impact of genetic engineering, provides most of the material for selective breeding of animals and plants. Not least, it is increasingly important for the understanding of human genetic disease, and is of inescapable though more controversial interest in connection with human diversity in general.

With the coming of DNA technology, the study of genetic variation has acquired a new dimension.

with the heterozygote; human sickle-cell anaemia, where the homozygote is sick but the heterozygote is resistant to malaria, is by far the best established example.

Secondly, in all outbreeding species, whether diploid or haploid, there is an abundance of variation detectable by electrophoresis of proteins or, still more, by restriction fragment analysis or sequence determination of DNA. This molecular variation shows perfectly clear-cut Mendelian inheritance but has, for the most part, little or no impact on the external phenotype or demonstrable effect on fitness. As we saw in the last chapter, this phenotypically cryptic variation provides most of the markers for the current projects for complete molecular mapping of the genome of various organisms.

Finally, again in outbreeding species, there is always a great deal of phenotypic variation of a kind that cannot be described in terms of clear-cut segregating differences but instead appears as a continuous gradation over a range. This is called *quantitative variation*.

Molecular identification of mutations of strong effect

The emphasis in this section is on examples from human molecular genetics, but exactly the same techniques can be applied to the identification of mutations in any organism for which the necessary background knowledge of chromosomes and linkage groups is available.

Using the mutation to find the gene

In order to determine the molecular nature of mutations it is necessary to clone the genes in which they occur. In micro-organisms there is often a good possibility of getting the gene to select itself, so to speak, through its ability to transform and 'rescue' a defective mutant (see pp. 84–85). Or, in any organism, if one or more proteins involved in the affected function are known, probes can be devised (p. 86) for the cloning of the corresponding genes, which can then be screened by one of the methods described below for mutations that might account for the observed phenotypes. This is the 'candidate gene'

approach. Frequently, however, the only way to find and clone a gene of unknown coding function and unclear affinities is through knowledge of its map position – *positional cloning*.

If a mutation of interest can be mapped between two molecular markers separated by a megabase or less, it should be possible to clone the whole sequence in between.

An outstanding success in clinical genetics was the identification of the gene involved in the common genetic disease, *cystic fibrosis* (CF) (Rommens *et al.*, 1989). Recessive defective CF alleles are present in as many as one in 25 of the UK population. At the start of the gene hunt there was no idea as to what the gene might look like, but it had been mapped approximately on chromosome 7 with reference to molecular markers. The region between two markers that appeared to bracket the gene was explored by a procedure known as *walking*.

Walking is the term applied to the step-by-step establishment of successive overlaps between cloned chromosome fragments, starting at a fixed point. Starting with a cloned fragment A, one screens for a second fragment B that cross-hybridizes and so overlaps with it, and then for a third fragment C that overlaps with B but not with A, and so on.

Chromosome walks can be thrown badly off track by repetitive DNA. Cross-hybridization between fragments will often be due not to a true overlap but rather to their both harbouring the same repetitive element or motif. Such sequences can sometimes be bypassed by a manoeuvre known as *jumping*. Briefly, the difficult fragment is cut out of the vector and circularized by ligation of its ends. A small fragment bridging between the original ends is obtained by digestion with another enzyme and cloned for use as a probe for the next step of the walk.

As more and more of the human genome is physically mapped by overlapping yeast artificial chromosome (YAC) clones, walking and jumping to find genes is becoming increasingly obsolete. Clones covering the region of interest will, to an increasing extent, already be available. But even if the desired gene is already present in a cloned fragment, there is still the problem of recognizing it when one sees it. In the case of the CF hunt, the responsible gene (or, rather, the first of its extremely widely spaced-out exons) turned up after only 300 kb of walking/jumping from one of the flanking molecular markers. It was identified by the following criteria.

1 It cross-hybridized to a sequence present in each of a number of different vertebrates. A positive result in such a test (sometimes called a 'zoo blot') is taken as evidence that the sequence is at least likely to be something important.

2 It hybridized to a human copy DNA (cDNA) clone corresponding to a messenger RNA (mRNA) that was synthesized particularly strongly in salivary gland tissue. This was very suggestive, since the salivary gland is one of the tissues in which the symptoms of cystic fibrosis are most evident.

3 Some 70% of CF patients were found to have a small deletion mutation at a particular point in one of the exons of the putative *CF* gene. Nearly 200 chromosomes not carrying a defective *CF* allele were screened for this mutation and none was found to carry it. As is so often the case, the mutation confirms the identification of the gene.

In some cases a mutant gene is marked more conspicuously, by a microscopically visible chromosome rearrangement. Whenever a chromosome is broken there is a chance of a gene being damaged at the break-point, either because of an interruption in its coding sequence or because of separation from an essential regulatory element. One example of the use of a segmental interchange to locate the gene responsible for a clinical trait is explained in Fig. 6.1. Here, a patient suffering from *neurofibromatosis* (NF), a condition of unknown molecular causation, was found to have a reciprocal exchange of segments of chromosomes 17 and 22. One of the break-points was near the centromere of chromosome 17, a region to which the *NF1* gene had already been approximately mapped by pedigree analysis. By isolating one of the interchanged chromosomes in a human–mouse hybrid cell line (cf. p. 145), and testing it for hybridization to a series of chromosome 17 DNA probes, it was possible to locate the break-point, and presumably the *NF1* gene, within a region bounded by two very closely linked molecular markers.

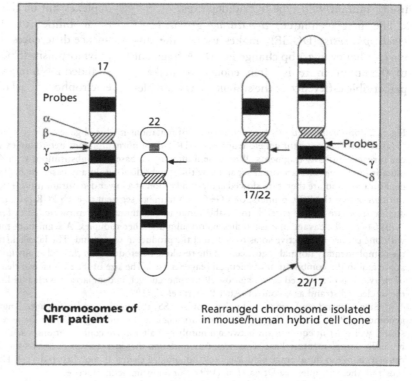

Fig. 6.1 An example of the use of a chromosomal segmental interchange for precise location of a gene. The exchange between the long arms of human chromosomes 17 and 22 was observed in a patient with the dominant condition neurofibromatosis (NF1), known to map to chromosome 17. Arrows indicated break–rejoin points. The break in 17 is presumed to disrupt the *NF1* gene. One of the interchanged chromosomes was isolated as the sole human component of the hybrid cell line and probed with a series of DNA probes, shown here as α, β, γ, δ, that identify four closely linked sites in the chromosome 17 physical map. Probes γ and δ hybridized and probes α and β did not, showing that the break was between β and γ. After Ledbetter *et al.* (1989).

Ways of screening known genes for sequence variation

Once a gene has been cloned and its sequence determined, it is relatively easy to find sequence variants, whether they have phenotypic effects or not. It is not necessary to clone the gene again from every individual one wishes to screen. With the wild-type sequence known, the polymerase chain reaction (PCR, see p. 147) can be used, with appropriate oligonucleotide primers, to amplify each new allele in segments from total genomic DNA for sequence comparison, segment-by-segment, with the wild type.

The work involved in searching a gene for a mutation can be greatly reduced if one knows more precisely where to look. In micro-organisms, or even in *Drosophila*, a fine-structure genetic map of mutational sites may be helpful. Such help is generally unavailable in higher organisms, but once the wild-type gene has been cloned and sequenced, there are several possible short cuts to the location of sequence variants. Three such methods are outlined in Fig. 6.2. All depend on comparisons of mutant and wild type gene fragments amplified by PCR, and the first two make use of mutant/wild-type hybrid duplexes.

The first technique, *denaturing gradient gel electrophoresis* (DGGE), makes use of the discovery that even a 1 bp change in a DNA fragment of 0.5 kb or more is often enough to make a perceptible difference to the stability of the double-stranded structure. If double-stranded fragments are run through an electrophoretic gel with an increasing gradient of urea and/or formamide (denaturing reagents that tend to 'melt' duplex to single strands), the fragments will run faster or slower according to their sizes until they reach the point in the gradient where their complementary strands start to come apart. At this point, their migration rate is drastically slowed, and fragments of even slightly different melting points will end up in distinctly different positions. A-T melt more easily than G-C base-pairs, and mismatches destabilize neighbouring base-pairs to different extents depending on the nature of the mismatch (Fig. 6.2a).

A second method, which has the merit of pin-pointing the position of a mutational difference within a fragment, depends on making mutant/wild-type hybrid duplexes and subjecting them to a chemical treatment that cleaves at mismatched pyrimidines – either at C or T depending on the chemical reagent used. Sizing of the cleaved fragments on a gel similar to that used for sequencing indicates where in the DNA fragment the mismatch occurred (Fig. 6.2b).

A third procedure depends not on making hybrid duplexes but on the distinguishable electrophoretic mobilities of DNA single strands with sequence differences – single-strand conformational polymorphism (SSCP). The rate at which single-stranded DNA migrates through a non-denaturing electrophoretic gel depends not only on its size but

Fig. 6.2 (*opposite*) Methods for identification of mutations in cloned genes.

(a) Denaturing gradient gel electrophoresis (DGGE) of normal/mutant heteroduplex made by hybridizing polymerase chain reaction (PCR) fragments. Where the mutation is a base-pair substitution, as in the example shown, the normal and mutant homoduplexes are often separable without hybridization. In any case, the heteroduplexes are more sensitive to denaturation and are therefore slowed in their migration at a lower denaturant (urea + formamide) concentration. Addition, through one of the PCR primers, of a G+C-rich terminal segment to each PCR product helps to maintain at least some duplex structure and so extends the usable range of denaturant concentration. After Higuchi *et al.* (1991).

(b) Chemical cleavage of mismatches in normal/mutant heteroduplex. A segment of the mutant allele is amplified by PCR, with one primer radioactive so as to end-label the product in one strand. The labelled strand is hybridized with an excess of the complementary normal sequence, and the resulting heteroduplex is cleaved at any mismatched pyrimidine – either C or T according to the combination of chemical reagents used. The size of the radioactive cleaved product, and hence the position of cleavage, is determined on an agarose electrophoretic gel, in comparison with a 'ladder' of fragments of known sizes and the uncleaved strand as a control. After Roberts *et al.* (1992).

(c) Single-strand conformational polymorphism (SSCP). Single strands of defined length isolated from normal and mutant genomes by PCR are run through a non-denaturing electrophoretic gel. Under the non-denaturing conditions, any single strand will exist in equilibrium between a number of alternative conformations, each held in place by annealing of short mutually complementary base sequences. Any mutation that increases or decreases the length of one of these sequences will cause a change in the average conformation and, often, a change in mobility in the gel. The examples shown are imaginary, to illustrate the principle. See Dryja *et al.* (1991) for an example in practice.

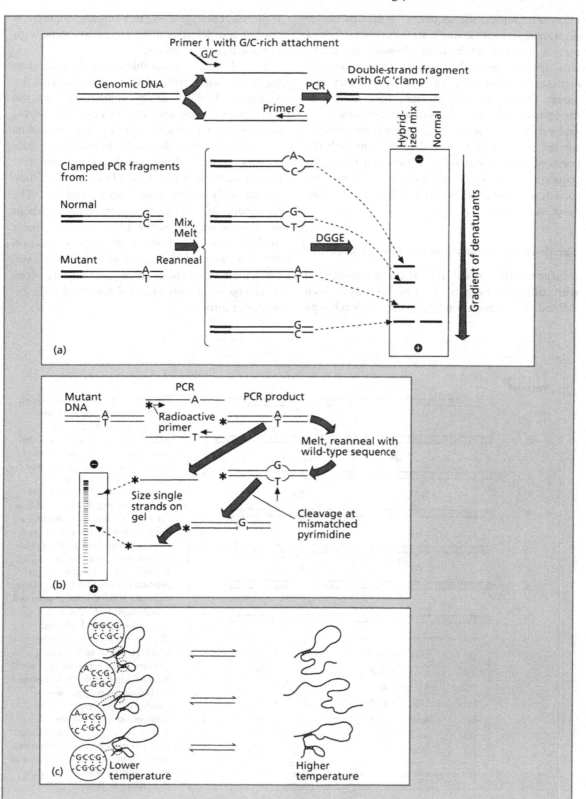

also on its nucleotide sequence. In the absence of denaturing reagents, single strands tend to form a certain amount of hydrogen-bonded secondary structure (stems and loops) by chance complementary matching between different short segments. Even single base changes will usually alter the relative stabilities of alternative conformations, and hence make it either easier or more difficult for the partly folded strand to pass through the pores of the gel (Fig. 6.2c). It turns out that most sequence variants in single-strand fragments of up to a few hundred bases in length can be distinguished in this way.

The diversity of mutant alleles in populations

A large number of mutational changes associated with different human diseases have now been defined. The general finding is that, for each type of disease, corresponding to a single gene, there is not just one defective allele but many, some with worse effects than others.

The *CF* gene is a notable example. Following the initial demonstration that about 70% of the defective *CF* alleles in the population were the same, with a 3-bp deletion in one of the exons, there was an intensive hunt for the mutations accounting for the other 30%. In little more than 2 years, about 170 individually uncommon mutations have been identified (Tsui, 1992). These include only a few more short deletions. The remainder are single base-pair changes; about 40% cause single amino acid substitutions, most of them very deleterious to the protein function but some with relatively mild effects. Other base-pair substitutions create chain-termination codons or change sequences essential for the proper removal of introns.

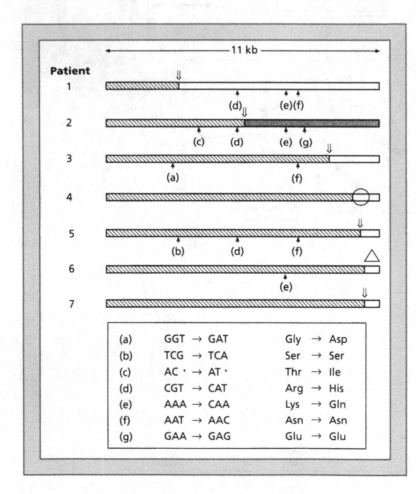

Fig. 6.3 Mutations found within the dystrophin gene in seven Duchenne muscular dystrophy patients. The diagram shows the positions of the changes in the 11-kb messenger RNA; in the gene itself the coding sequence is dispersed over 79 exons. In the cases shown the disease condition can be attributed to truncation of the polypeptide chain from the C-terminal (right-hand) end, due to chain-termination mutations (⇓), defective intron splicing (○) or deletion of exons (△). The relative mildness of the disease in patient 2 was thought to be due to an alternative mode of splicing, effectively bypassing the premature termination site and allowing formation of a partially functional product (stippled bar). Also found in these seven patients were a number of apparently inconsequential sequence variations (a–g); several of these recurred in more than one patient and were presumed to be common in the normal population. After Roberts *et al.* (1992).

The intensive work on cystic fibrosis has provided more information on relative frequencies of defective alleles than is available for any other system. One allele accounts for about 70% of the total, about 20 others are comparatively common, together bringing the proportion up to about 90%, and the remaining 150 or so are very rare, many having only been recorded once. Why some defective alleles should be so much commoner than others is an interesting question, perhaps answerable in terms of genetic drift and founder effects when populations expand from initially small numbers – topics beyond the scope of this book.

Another human gene that has been intensively searched for mutations is that encoding dystrophin, the protein that is defective in the sex-linked disease Duchenne muscular dystrophy. Here about two-thirds of the disease alleles have fairly extensive deletions, presumably because relatively few single amino acid changes in the enormously long coding sequence (11 kb, distributed between 79 different exons) injure function to the extent of causing clinical symptoms. The mutations of serious effect that are not deletions mostly cause chain termination or, in one case, a defect in intron splicing. Interestingly, in the course of the identification of the crippling mutations, a number of other base-pair changes were found that were not obviously relevant to the disease phenotype (Fig. 6.3). Some of these were changes in third positions of codons that merely changed one codon for another encoding the same amino acid – a 'silent' or synonymous change – but some caused amino acid replacements. Since several of these variants were found repeatedly in patients with different mutations of severe effect, it was presumed that they were polymorphisms present in the normal population. We return below to the question of 'neutral' variation in protein sequences.

The kinds of mutations likely to be found in functionally defective alleles clearly depend on the nature of the protein gene products. Some are sensitive to many single amino acid replacements, while others may be robust to most changes short of losses of substantial lengths of polypeptide chain. But among all mutations, both severe and mild in effect, single base-pair substitution are the most frequent. Within this class, there is, in

humans at least, a substantial bias in favour of transitions from G-C to A-T. Cytosines immediately 5′ to guanines seem to be particularly vulnerable to mutation. In mammals, such Cs are prone to methylation to 5-methyl-cytosine, which may be spontaneously deaminated to form uracil and an A-T base pair after the next round of replication (see p. 156).

Screening for known mutations

When a mutant allele of a particular gene has been defined in terms of DNA sequence, a specific *allele-specific oligonucleotide*, or ASO, can be used to probe for its presence in any individual genome. An ASO is a synthetic sequence of about 18 nucleotides made to match one DNA strand of the mutant allele, with the site of the mutation centrally placed within it. When used as a radioactive probe on membrane-bound DNA samples, it will, at the right temperature and salt concentration, hybridize very much better to the mutant than to the normal sequence (Fig. 6.4).

Comprehensive screening of human populations for carriers of a particular kind of recessive condition is generally thought to be worthwhile only when a very high proportion of defective alleles can be identified. At present, it is likely that 10% or more of cystic fibrosis carriers would remain undetected with the probes currently available.

DNA sequence variation without phenotypic consequences

The scope for neutral mutation in protein-encoding genes

The ability of many enzymes to maintain their functions in spite of amino acid replacements has long been apparent from studies on microorganisms. When selection has been made for mutations restoring normal enzyme activity to an auxotrophic mutant of a bacterium or fungus, the restored enzyme is not always exactly the same as in the original wild type. The mutant codon, which in different cases may be a chain-terminator or one encoding an unacceptable amino acid, can be converted by a single base mutation to nine other codons, several of which, encoding different

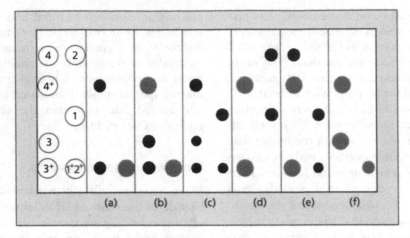

Fig. 6.4 The use of allele-specific probes for the identification of cystic fibrosis (CF) mutations in humans. The 18-nucleotide synthetic probes were made to match the commonest mutant sequence (shown here as 1), three less common mutants (2, 3 and 4), and the corresponding wild-type sequences (1^+–4^+). 1 and 2 were different deletions in the same sequence, so the same normal control (1^+2^+) serves for both. The probes were bound in a standard pattern (left-hand panel) to nylon membranes, which were then hybridized to appropriate exon sequences amplified (and radioactively labelled) by PCR from the DNA of six people (a–f): (a) normal; (b) normal but father of (c), and (c–f) all cystic fibrosis patients. The panels (b–f) represent the autoradiographs. The results identify CF mutations as follows: (a) none detected; (b) 3, together with a normal allele; (c) *1/3*; (d) *1/2*; (e) *1/1*; (f) *3/3*. Note that the probe for 4, which is a single base-pair replacement, may give some signal with the normal sequence, though not strong enough to cause ambiguity. After Serre *et al.* (1991).

amino acids, may be compatible with normal function – or at least sufficiently normal not to cause any evident phenotypic abnormality. At many positions within enzyme polypeptide chains, conservative amino acid substitutions – those that replace one residue by another of similar properties, for example isoleucine for valine or aspartate for glutamate – have little or no obvious effect on growth rate. At some positions even non-conservative changes have minimal effect. But many changes that have little or no effect on function as measured in the laboratory may well be of consequence in the real world.

In humans, screening for mutations in genes at the DNA level has been largely confined to alleles known to be defective. Most often, only one mutation has been found distinguishing a given defective allele from the common normal sequence. For many genes there does indeed seem to be a standard 'wild type' so far as protein-coding sequence is concerned. However, as Fig. 6.3 shows, this is not the case for all genes. And even before the advent of DNA sequencing, the existence of multiple functionally normal alleles of a number of human genes had been revealed by

electrophoretic differences between their enzyme products.

Some of the most extensive surveys of allelic variation using enzyme electrophoresis have been made in *Drosophila* species. Figure 6.5 shows one example. To find only a single wild-type allele by this criterion is more the exception than the rule. The *Drosophila pseudoobscura* gene encoding the enzyme xanthine dehydrogenase is a striking, perhaps rather extreme, example. Here the use of a number of different conditions of electrophoresis, thought to reveal changes in protein shape as well as in electrical charge, allowed the identification of 37 different alleles in a sample of 146. All the electrophoretic variants seemed quite adequately functional and there was no reason to suppose that they were associated with any differences in fitness of the organism. It is not possible to point to a single standard wild type.

Another easily scored *Drosophila* enzyme, alcohol dehydrogenase (ADH), seems much less prone to apparently neutral variation in amino acid sequence. In *D. melanogaster*, there are two common electrophoretically distinguishable forms, apparently maintained for unknown rea-

Fig. 6.5 Variation in the enzyme esterase-5 in two populations of *Drosophila pseudoobscura*. Protein extracts from individual flies were fractionated by electrophoresis, and the gels were stained with a reagent that is converted to a dye by action of the enzyme. The sample shown contains five flies from each of two populations with two from a standard laboratory strain (S) for comparison. Enzyme variants correspond to three different equally functional alleles of the *Est5* gene (represented here by *a*, *b* and *c*), two of them common in each population. Heterozygotes show three-banded patterns because the enzyme is a dimer of two polypeptide chains, both encoded by *Est5*, so that a proportion of hybrid dimers is formed when two different alleles are present. In the *b/c* heterozygote the bands are incompletely separated. After Hubby & Lewontin (1966).

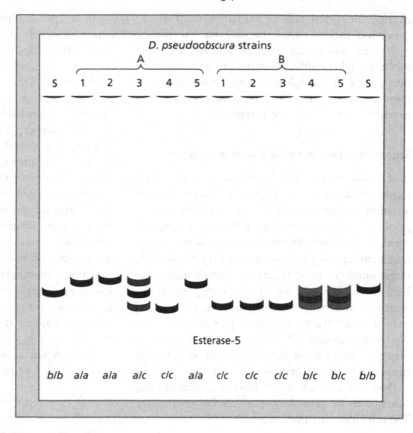

Table 6.1 Variation within the *Adh* gene of *Drosophila melanogaster*. Complete sequences were obtained from 11 chromosomes collected from five populations from different parts of the world. Data from Kreitman (1983)

	5'-flanking	5' untranslated exon	Introns	Amino acid-encoding	3' untranslated exon	3'-flanking
Total base-pairs	63	157	755	768	178	767
Insertions/deletions	0	0	2	0	1	3
Variable bases	3	1	20	13	2	5
Amino acid replacement	0	0	0	1*	0	0

* This is the common fast/slow electrophoretic polymorphism (aspartate/threonine) found in nearly all populations; 12 out of 13 of the changes observed in codons were 'silent'. Note that 75% of random base substitutions in codons would give amino acid replacement.

sons as a stable polymorphism, but otherwise very little overt variation. However, DNA sequencing of a sample of wild-type *Adh* alleles revealed many cryptic differences (Table 6.1). There were numerous base substitutions in the third positions of codons, as well as in the introns and sequences flanking the gene. These results fall into a general pattern that has been repeatedly recorded in com-

parisons between related species in many groups of plants and animals. Differences in codons that affect the nature of the amino acid encoded are far less frequent than synonymous changes, but they do occur. Differences in introns, when outside the consensus sequences essential for splicing, occur much more freely.

It may be questioned whether synonymous

changes in codons are necessarily neutral. If the alteration in the codon necessitated the use of a different and scarce transfer RNA (tRNA) molecule, it might be difficult to maintain a translation rate sufficient for fulfilment of normal function. However, no overt effects of synonymous codon changes seem to have been reported.

Sequence variation outside coding sequences

In the genome as a whole, including all the DNA sequence that lacks clear specific function, there is an immense amount of cryptic variation in any population that is not highly inbred or propagated clonally. We saw in the last chapter how restriction fragment-length polymorphisms (RFLPs) provide large numbers of markers for genetic mapping. RFLPs are especially good for detecting changes in numbers of repetitive sequences, but they reveal only a small fraction of base-pair substitutions. For every sequence difference that affects a restriction site, there must be many more that do not draw attention to themselves in this way.

Does all this variation have any significance, other than in providing markers for geneticists? We return to this question later.

Quantitative variation and its heritability

Means and variances

If one cannot describe variation in terms of sharply distinct phenotypic classes there is no alternative to doing it by measurement. Often, the feature under study is essentially quantitative in the first instance, as with height, weight or (a favourite of *Drosophila* geneticists) number of bristles. In other cases an artificial scale of measurement has to be devised, even at the cost of some degree of over-simplification. The measurement of human intelligence in terms of IQ is an example.

Variation within a population of organisms with respect to a quantitative character very often approximates to a *normal distribution* about a mean. Without giving the mathematical formula for the normal distribution, it can be described as what one would expect if deviations from the mean were due to a large number of relatively small independent perturbing influences, each one acting positively or negatively with equal probability. If the positive and negative perturbations happen to balance the result will be a value equal to the mean. This, statistically, will be the most probable outcome. Deviations from the mean become less probable the more extreme they get, because they depend on increasing numbers of independent perturbations just happening to act in the same direction. The curve representing the distribution of values about the mean is symmetrical and bell-shaped, as shown in Fig. 6.6. The degree of spread of the curve is expressed as the *variance*, which is the sum of squares of the deviations from the mean divided by one less than the number of observations.

If the n members of the population have the values $x_1, x_2, x_3 \ldots x_n$, the mean of the population is $(x_1 + x_2 + x_3 + \ldots + x_n)/n = \bar{x}$.

The individual deviations from the mean are $x_1 - \bar{x}, x_2 - \bar{x}, x_3 - \bar{x}$ etc.

The variance is sum of the squares of the deviations divided by the total number minus 1:

$$V = \sum_{1}^{n} (x - \bar{x})/(n - 1)$$

The reason why the sum of the squared deviations is divided by $n - 1$ rather than by n is that the amount of deviation from the mean has to be assessed in relation to the number of opportunities for deviation, or the number of *degrees of freedom*. With two measurements, one can vary from the mean and then the other is fixed; with three, two can vary independently and then the third is fixed, and so on. One degree of freedom is always used up in the calculation of the mean.

The standard deviation from the mean (σ) is the square root of the variance:

$$\sigma = \sqrt{V}$$

The standard deviation expresses the amount of variation from the mean in the same units as the measurements themselves. It can be related directly to the normal distribution, being the distance from the mean to the point of inflection of the curve on each side. It is useful to remember that the proportion of a normally distributed population falling more than two standard de-

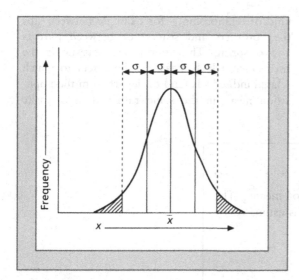

Fig. 6.6 The normal distribution generally exhibited by quantitatively varying traits in populations. It is what is expected if deviations from the mean are due to the combination of numerous independent effects of similar magnitude. The frequency of different values of the variable x is distributed symmetrically about the mean value, $\bar{x}$. The standard deviation σ, the square root of the variance, is the distance from the mean to the steepest point on the curve (the point of inflection) on each side of the mean. The proportion of the population deviating from the mean by more than 2σ is approximately 5%.

The fact that variance is additive means that it is possible in principle to partition the total phenotypic variance of a population (V_p) into components due respectively to environmental differences (V_e) and genetic differences (V_g).

$$V_p = V_g + V_e$$

The first step in quantitative genetic analysis is to determine whether the varying trait is heritable at all – whether, in other words, the variance has a genetic component. In its broad meaning, heritability is that proportion of the total variance that is due to genetic factors:

Heritability (broad sense) = V_g/V_p

In its stricter (narrow) sense, heritability is a measure of the actual or hypothetical response to selection. If one selects as parents of the next generation individuals measuring a certain amount, call it S, above the population mean, and their progeny average R above the mean, then the narrow-sense heritability, conventionally called h^2, is equal to R/S.

If all alleles affecting a metric character were simply additive in effect, then h^2 would be equal to V_g/V_p, that is the broad-sense heritability. Putting it another way, all the genetic variance would be selectable. In fact it is not, the most important reason being that some of it is due to dominance relationships between alleles – their non-additive effects in heterozygotes.

On certain simplifying assumptions, V_g can be partitioned into additive and dominance components, $V_g = V_a + V_d$, and heritability in the strict sense is not V_g/V_p but V_a/V_p, the fraction of total variance that can be fixed by selection.

viations from the mean (the hatched part of the area under the curve in Fig. 6.6) is approximately 5%.

The usefulness of the variance in statistical calculations lies in the fact that it is an *additive* measure. It increases in direct proportion to the number of perturbing influences, assuming them all to be of roughly similar magnitude. The standard deviation, on the other hand, increases only in proportion to the statistical probability of random perturbations reinforcing each other as opposed to cancelling out, and this is proportional to the square root of the number of perturbing influences.

Heritability of quantitative variation

Some quantitative variation is due to environmental fluctuation and is not heritable, but in outbred populations a high proportion of it can be shown to be genetically based.

Heritability assessed by correlations between relatives

There are two general appproaches to the analysis of genetic variance. One starts with a cross between two phenotypically different inbred strains of a species, and proceeds through the calculation of means and variances from F_1, F_2 and backcross progenies. This gives information only about the differences between two arbitrarily selected homozygous genotypes, such as can be obtained in self-fertilizing plants or laboratory stocks of

Drosophila or mice, and depends on the availability of large progenies. A more generally useful approach is through calculations of correlations between relatives.

Suppose one wants to calculate the degree of phenotypic correlation associated with a certain kind of relationship – for example between sibs (i.e., brothers and sisters) or between parents and offspring. The appropriate statistic is the *covariance*, which measures the extent to which related individuals tend to depart from the population mean in the same rather than opposite

---

Box 6.1 Covariance and correlation

· ·

Consider a quantitative character x, with population mean x. The variance of the population (P) with respect to x is estimated by the formula:

$$V_P = \sum_1^n (x - \bar{x})^2 / (n - 1)$$

where n is the size of a random population sample.

Now consider pairs of individuals that bear a certain relationship – say parents and offspring, sibs, half-sibs. Denoting one side of the relationship as a and the other as b, the covariance of groups a and b is given by the formula:

$$W_{ab} = \sum_1^n (x_a - \bar{x})(x_b - \bar{x}) / (n - 1)$$

where n is the number of $a - b$ pairs. We assume for simplicity that each group has the same mean and variance as the population as a whole.

Correlation coefficient, $t = W_{ab} / V_P$

If the two groups are perfectly correlated, so that a given a and b always have identical deviations, positive or negative, from the population mean, the product of their deviations will be the same as their squares, so

$W_{ab} = V_P$ and $t = 1$

If the two groups are totally uncorrelated, the members of a pair will deviate in opposite directions from the mean as often as in the same direction, the products of their deviations will be as often negative as positive, and W_{ab} and t will both be zero.

If the two groups a and b have different means and variances, for example because they come from different generations with different environments, then:

$$W_{ab} = \sum_1^n (x_a - \bar{x}_a)(x_b - \bar{x}_b) / (n - 1) \quad \text{and} \quad t = W_{ab} / \sqrt{V_a V_b}$$

directions. The method of calculation is summarized in Box 6.1.

If the relatives were perfectly correlated – that is, if their deviations from the mean were always identical – their covariance would be equal to the population variance. If, at the other extreme, there were no correlation at all, a positive deviation from the mean on the part of one relative would as often as not be accompanied by a negative deviation on the part of the other. The products of the deviations would be as often negative as positive and the summed products would approximate to zero. The ratio of the covariance to the overall population variance is called the *correlation coefficient*; it is zero for no correlation and unity for complete correlation.

Covariance of relatives is an expression of the extent to which they share the same components of variance. There will be an environental component to be taken into account if relatives tend to have more similar environments than unrelated individuals – we can call this *environmental correlation*. It is always likely to be important in human populations. For farm animals kept under a common regime the environmental correlation component of variance (V_{ec}) is arguably significant only in the case of sibs. In this case covariance (W) can be related to components of variance as follows:

parent-offspring	$W_{PO} = V_a/2$	(1)
full sibs	$W_{FS} = V_a/2 + V_d/4 + V_{ec}$	(2)
half-sibs	$W_{HS} = V_a/4$	(3)
identical twins	$W_{mz} = V_a + V_d + V_{ec}$	(4)

The reasons for the different fractions of V_a should be obvious. Offspring inherit a half of their alleles from each parent. Sibs have a 50% chance of sharing any particular parental allele, and half-sibs a 25% chance. The dominance component V_d can only contribute to resemblance between relatives if, because of their relationship, they can have both alleles of a particular gene in common. This can only apply to sibs, and the chance of sibs both having the same allele from one parent and both the same allele from the other is $1/2 \times 1/2$ or $1/4$. Identical twins (*monozygotic*, derived from the same fertilized egg) have identical genotypes.

Putting together the data from a number of different correlations between relatives, it is poss-

ible to estimate both V_a and V_d and, by comparison with the total population variance, heritabilities in both the broad and narrow senses.

Estimates of heritability in humans

The confusing effect of environmental correlation makes heritability much more difficult to calculate for human populations. The only plausible way of doing it is by the use of twins. Many twin studies have been carried out, both with monozygotic and dizygotic twins, especially in connection with the heritability of intelligence as measured by IQ.

Dizygotic twins, being the products of separate egg–sperm unions, have the same average degree of genetic similarity as pairs of non-twin sibs, and Equation 2 above again applies. Assuming that the environmental correlation is the same for monozygotic (mz) as for dizygotic (dz) pairs, we can substract Equation 2 from Equation 4 to obtain:

$$W_{mz} - W_{dz} = V_a/2 + 3V_d/4$$

There is no way here of disentangling V_a from V_d, but the latter term is likely to be relatively small. To a reasonable approximation, the difference between the two covariances can be taken as equal to half the total genetic variance V_g.

Table 6.2 shows correlations (covariance ÷ population variance) between ordinary sibs and the two kinds of twins with respect to finger-ridge

Table 6.2 Estimated heritabilities of some human traits. From Falconer (1989)

	Finger-ridge count	Height	IQ
Correlations (t)			
Monozygotic twins	0.96	0.90	0.83
Dizygotic twins	0.47	0.57	0.66
Full sibs	0.51	0.50	0.58
Heritability			
2 (tFS)*	1.02	1.00	1.16
2 (tMZ-tDZ)†	0.98	0.66	0.34

* The unreal heritability estimates of more than 1 are due to lack of correction for environmental similarity.
† The substraction of the dizygotic from the monozygotic twin correlation supposedly corrects for environmental similarity.

count, height and IQ. Also shown are the heritabilities calculated on the twin data (as just described) and on sib data by ignoring both V_d and V_{ec} and just doubling the sib covariances (see Equation 2) Finger-ridge count appears to be totally heritable, and the other two characters partially heritable. For both height and IQ the 'heritability' estimates from the ordinary sibs, in which environmental correlation (V_{ec}) is ignored, are considerably greater than from the twin calculation, in which V_{ec} is supposedly cancelled out. This illustrates the potency, with regard to these characters, of environment differences. Nevertheless, the twin calculation still indicates a substantial role for inheritance in the determination of both characters.

Another way of trying to eliminate the effect of environmental correlation is to base the heritability estimates on pairs of identical twins reared apart. Several such studies have been made in relation to IQ, and they have all shown very significant heritability – generally greater, indeed, than the estimate shown in Table 6.2.

Questions have still been raised as to whether environmental correlation has really been eliminated. There could be a significantly greater similarity between the environments of monozygotic than of dizygotic twins, since each twin is an important part of the environment of the other. It has also been suggested that selective placing by adoption agencies could result in a significant degree of environmental correlation between twins reared apart; it is difficult to see, however, how any such effect could explain the greater correlation of monozygotic than of dizygotic twins.

How many genes?

It needs to be demonstrated that heritable differences with respect to quantitative traits are attributable to genes segregating in meiosis according to the Mendelian rules. The simplest argument rests on the determination of means and variances of F_1 and F_2 progenies of crosses between inbred strains of plants or animals. If the parental strains are inbred to the extent of becoming homozygous with respect to every gene, they will each display only environmentally-caused variation. The F_1 progeny will likewise be genetically homogeneous,

since each individual will be heterozygous with respect to any pair of alleles in which the parents differ. One might expect that the parental and F_1 variances will all be similar – V_e with no V_g component – though, strictly speaking, the three populations might have somewhat different values of V_e because of different degrees of sensitivity to environmental variables. In any case, the F_2 generation always shows a greatly increased phenotypic variance. This is what would be expected from simple gene theory, since each gene difference A/a contributing to the difference between the parental means will show a $1:2:1$ $AA/Aa/aa$ ratio. The effects of a number of independent $1:2:1$ ratios, when superimposed, will be to generate additional genetically based variation, broadening the bell-shaped normal curve. If the variation was attributable to only one or two allelic pairs one would expect a stepped rather than a smooth distribution, and dominance of an allele of major effect could skew the curve. But in practice, given that each genotype will be subject to its own environmental variance it requires only a moderate number of gene differences for the steps to be smoothed out, and dominance effects, if any, tend to balance out overall (Fig. 6.7). So one tends to get a reasonably normal distribution in the F_2. The F_2 variance consists of the V_e component, which can for the purpose of argument be equated to the parental and F_1 variances (or their average if they differ), plus a V_g component, which can be obtained by subtraction.

With the aid of a number of simplifying and somewhat implausible assumptions, it is possible to use F_2 variance to calculate the minimum number of gene differences contributing to the quantitative variation. The fewer the differences, the greater one expects the F_2 variance to be in proportion to the difference between the parental means. The great American quantitative geneticist, Sewall Wright, derived the following formula for the number of effective factors, n_E:

$$n_E = (\mu_1 - \mu_2)^2/8V_g,$$

where μ_1 and μ_2 are the parental means and V_g is the genetic variance of the F_2, obtained by substracting V_e from the total phenotypic variance. The number of effective factors is the number of gene differences that could account for the

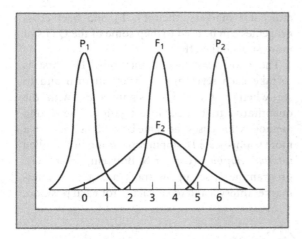

Fig. 6.7 Normal distributions in parental, F_1 and F_2 generations in the hypothetical case that two inbred parent lines have genotypes *a/a b/b c/c* and *A/A B/B C/C* with each capital-letter allele contributing +1 and each lower-case allele contributing 0 to a quantitative trait. The three genes are unlinked and there is no dominance. The parental means are 0 and 6 and the F_1 mean 3 (*A/a B/b C/c*). In the F_2 there are 64 random combinations of the eight gamete genotypes, generating the values 0, 1, 3, 4, 5, 6 in the ratio $1:6:15:20:15:6:1$. If each of these F_2 classes has the same environmental variance as shown by the parental and F_1 generations, this stepped distribution will be smoothed to approximate to the curve shown. Substracting the parental/F_1 variance from the F_2 variance gives the genetic component V_g of the F_2 variance. In this artificially simple case the difference between the parents is 6 and V_g comes out to be 1.5; $6^2/(8 \times 1.5) = 3$, the number of gene differences (Wright's formula, see p. 176).

observed F_2 variance if the assumptions underlying the calculation were all true.

These assumptions are: (i) no linkage; (ii) no dominance (i.e., heterozygous loci always contributing a value midway between the homozygotes); (iii) the differences between the parental lines all act in the same direction, i.e., all positive in one and all negative in the other; and (iv) all differences are of equal effect on the phenotype. All are manifestly unrealistic, and the more wrong they are, the greater will be the underestimation of the number of gene differences. Regarding assumption (iv), if gene differences are having a range of effects, going all the way down to zero, they could be indefinitely numerous. Since we know that linkage does exist, and the number of chromosomes is limited, the effective factors can more realistically be equated to the segments, each

containing at least one relevant gene, into which the chromosomes are divided by crossing over.

In spite of these reservations, some conclusions can be drawn. It can be shown that even if the effects of different genes are unequal, there must be at least one effective factor that makes a contribution of at least $(\mu_1 - \mu_2)/n_E$, the calculated average effect. Applied to a range of organisms, including animals, plants and humans, the method, with some recent refinements (Zeng, 1992), has generally yielded values of n of between 5 and 20, which at least suggests that the gene differences are not all so small in effect as to discourage attempts to identify them individually.

What kinds of genes are involved in continuous variation?

The total effect of a gene cannot be inferred from the effect of any particular mutation that occurs within it. It may well be that most of the gene differences underlying quantitative variation are minor variations in genes with major functions, complete elimination of which would have drastic effects. Some of the variation that has been detected within the amino acid sequences of essential proteins, and which at first sight seems 'neutral', may well have small effects. Variation in promoters and enhancers could result in small as well as large quantitative differences.

An older idea, embodied in the term *polygene*, is that the genes involved in quantitative variation are a special category, involved more in 'fine-tuning' of the phenotype than in the provision of essential and unique functions. Highly repetitive genes, such as those for ribosomal RNA (rRNA) or histones, could be polygenes in this sense. So, too, might be some of the genes which are being identified in genome projects as apparently redundant open reading frames (see pp. 157 and 215). It is difficult to be sure, however, that an apparently near-inconsequential gene might not have a crucial function under conditions not yet tested.

The problem of identifying gene differences of individually indistinguishable effect is always going to be intractable. The hope is to reduce this problem to a minimum by accounting for as much variation as possible in terms of measurable differences due to identifiable gene loci.

Attributing heritable quantitative variation to chromosomal loci

Quantitative trait loci distinguishing inbred strains

It is the ambition of many plant and animal breeders to pin down the determinants of quantitative variation to particular points on the chromosomes – *quantitative trait loci* or QTLs. If quantitatively different inbred strains are distinguished by a sufficiently large number of mapped molecular markers, their F_1, F_2 and back-cross progenies can be used to map some of the QTLs of more substantial effect.

There are two general methods. The first is to take each segregating marker in turn and to see whether it correlates significantly with the quantitative trait under investigation. The second method, which uses the same basic data but with a more sophisticated computer programme, is called *interval mapping* (Lander & Botstein, 1989). The programme is set up to track along each inter-marker interval and to calculate, for each position,

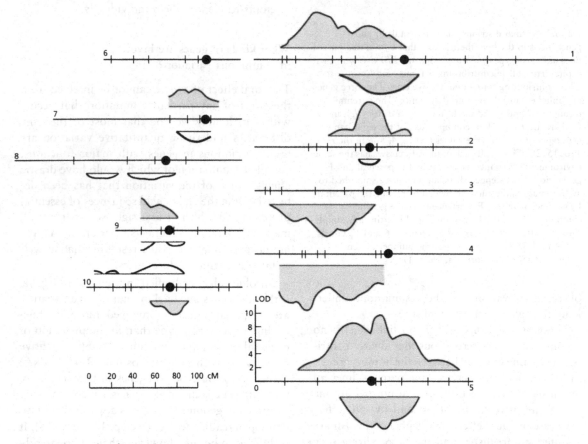

Fig. 6.8 The results of interval mapping of the quantitative trait loci (QTLs) involved in the determination of the heterosis in a maize hybrid. Two 'elite' inbred lines were crossed: F_2 plants were self-pollinated to generate 264 different F_3 progenies. A single plant from each F_3 was typed with respect to 76 different molecular markers differentiating the two original inbreds. These F_3 plants were back-crossed to each of the two inbreds and the grain yields of the progenies (nearly 100 000 plants in all) were determined. The data were processed by computer to generate the LOD curves shown. The positions of the markers are shown on linkage maps representing the 10 chromosomes (centromeres shown as filled circles). The curves, which are cut off below LOD = 2 (deemed to be the level of significance), show the distribution of likelihood of the presence of a QTL, derived from one inbred line, conferring heterotic vigour in the genetic background of the other. The two different sets of curves, shown respectively above and below the linkage maps, represent the results from back-crosses to the two different parental lines. After Stuber *et al.* (1992).

the probability of getting the observed data with a QTL in that position divided by the probability if there were no QTL there. The ratio is expressed as the logarithm to give a LOD score, closely analogous to the LOD measure used to map clear-cut markers in human genetics (see p. 28). Some arbitrary LOD score, say 2, has to be taken as a cut-off point below which the evidence for the presence of a QTL is deemed to be too weak to be worth pursuing. Different computer programmes have been devised for dealing with different kinds of progeny – F_2, F_3 or back-cross.

The advantage of interval mapping as compared with correlating the trait with all the markers one-by-one is that it gives a more precise indication of positions within intervals. It also gives a better estimate of the contribution of each identified QTL to the total quantitative difference; correlation with just a single marker does not discriminate between a relatively weak effect with close linkage and a stronger effect with loose linkage.

An example of the use of interval mapping to map QTLs concerned with grain yield in maize is shown in Fig. 6.8. Commercial maize production is based almost entirely on hybrids between selected inbreds. The advantage of hybrids is their uniformity and high yield. The fact that they perform much better than the parental inbreds – the phenomenon of *heterosis* – implies either that heterozygosity is advantageous as such or, more likely, that a certain number of somewhat deleterious alleles inevitably become fixed in the process of inbreeding. In the study in question, 264 F_3 plants stemming from a cross between two commercial inbreds were each back-crossed to both parental lines. The F_3 plants were all scored with respect to over 79 molecular markers distinguishing the parent inbreds. Thus each F_3 plant was known to be contributing, in the back-cross to one parent, a particular defined set of marked chromosome segments derived from the other. The interval analysis was designed to reveal the whereabouts of the QTLs in each parent line that contributed to heterosis in combination with the genome of the other line. These different contributions were reflected in the grain yields of the back-cross progenies, all 528 of which were grown under controlled field conditions.

As Fig. 6.8 shows, QTLs of significant effect were identified on nine of the 10 chromosomes. Those of strongest effect contributed as much as 15–20% of the total, and others were only barely significant at the LOD = 2 level.

Although successful up to a point, this kind of analysis is apt to be discouraging if the ultimate objective is to clone and characterize the genes at the QTLs. The indications of position in Fig. 6.8 are too imprecise to provide starting points for molecular identification of the genes concerned. The position could be improved by using an even larger number of molecular markers at closer spacing, but, even so, it will always be difficult to find genes just by interval mapping without further clues.

Another example of interval mapping shows some of the ways in which other clues can be obtained. It concerns genetic variation in blood pressure in rats; two inbred strains were found to differ substantially in this quantitative trait. The molecular markers required for analysis in terms of QTLs were provided by using PCR to amplify a large number of short genomic segments containing simple-sequence repeats (cf. p. 132). Used as probes, these PCR products revealed a total of 112 sequence-length differences between the two strains, and these were used as molecular markers for analysis of the F_2 generation (115 animals) from the inter-strain cross. All the F_2 rats were scored with respect to all 112 markers. On the basis of the results the markers could be assembled into linkage groups, most of which were shown by hybrid (rat–mouse) cell analysis (see p. 145) to correspond to known chromosomes. The F_2 rats were also scored for blood pressure. The most striking out-come of the quantitative analysis was that nearly 20% of the variance in the F_2 with respect to blood pressure could be attributed to a QTL on the 10th chromosome, which had been divided into four intervals by five molecular markers. The result of the LOD analysis is shown in Fig. 6.9. It turned out that the most likely place for the chromosome 10 QTL was inseparable from one of the markers, RD17. Given the numbers analysed, this was still consistent with the QTL and RD17 being separated by a few centimorgans and a few million base-pairs.

As it turned out, the marker itself provided a

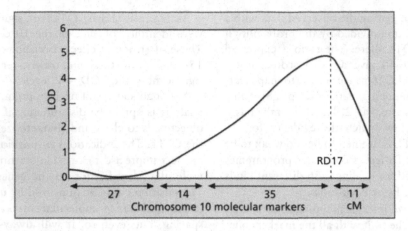

Fig. 6.9 Interval mapping of rat chromosome 10 for the location of quantitative trait loci (QTLs) concerned with blood pressure. Two inbred lines strikingly different in blood pressure were typed with respect to a number of molecular markers, five of them on chromosome 10. The LOD curve shown was generated by analysis of F_2 mice from the inter-strain cross. The most likely position for a chromosome 10 QTL turned out to be virtually inseparable from the marker RD17. It accounted for approximately 20% of the total F_2 blood pressure variance. After Jacob *et al.* (1991).

vital clue in this case. It was a polymorphism found by use of a probe for the gene encoding growth hormone (GH). It was also known that there was close similarity between the arrangement of genes in that part of rat chromosome 10 and a corresponding part of human chromosome 17 harbouring the human GH gene. Very closely linked to the human GH gene is a gene encoding angiotensin-converting enzyme (ACE) which is known to have an important role in blood pressure control. The strong hypothesis, which should not be hard to check, is that it is variation in the rat ACE gene that is responsible for a part of the inter-strain blood pressure difference. This, if true, may help the understanding of congenital variation in human blood pressure also (Leckie, 1992).

The foregoing example shows the usefulness of inter-species comparisons in gene identification. Different mammals have different chromosome numbers and different combinations of chromosome segments, but within a segment the arrangement of genes is often remarkably similar. The same undoubtedly applies to other groups of organisms. The other lesson one can draw from this example is the importance of good luck in finding close and informative linkages.

Looking for QTLs in outbred populations

The type of analysis described in the preceding section depends absolutely on the molecular polymorphisms being reliable tags for the QTLs. This is a safe assumption when one starts with highly inbred lines, with all individuals virtually identical and homozygous at all loci. In larger farm animals, and, of course, in human populations, this is far from being the case. Many individuals are heterozygous both for the QTLs that one would like to map and for the linked molecular markers. At every generation, recombination can occur to generate new combinations of quantitative alleles and the markers linked to them. In time, longer for closer linkages, a situation of *linkage equilibrium* will be reached.

Suppose one has two QTL alleles a_1 and a_2 with frequencies p and $1 - p$ in the population, and a linked marker locus with alleles m_1 and m_2 at frequencies q and $1 - q$. It might be that marker allele m_2 originally arose by mutation in a chromosome carrying a_2. If the linkage was fairly tight, m_2 could remain a reasonably reliable marker for a_2 for a considerable number of generations. But eventually, and assuming no significant effect of selection, the population would come to equilibrium with $a_1\,m_1$, $a_2\,m_2$, $a_1\,m_2$ and $a_2\,m_1$ chromosomes at frequencies pq, $(1-p)(1-q)$, $p(1-q)$, $(1-p)q$. At this point, random mating and recombination will be restoring and disrupting the original $a_2\,m_2$ linkage at equal rates. In this situation of linkage equilibrium the marker will be no use at all as a tag for the QTL.

The degree of genetic disequibrium, and hence the reliability of a marker, will decline over generations, but it may be increased by extreme fluctuations in population size. If there have been population 'bottlenecks', with temporary reductions to relatively few individuals, it is likely that some linkage disequilibrium will arise by chance. For example, to take our previous model, it could be that the few surviving a_2 chromosomes all happened to carry m_1 This sort of effect (called 'founder effect' in population genetics) is likely to be important in pedigree breeding of animals.

Reasonably consistent associations between markers and QTLs cannot be expected if the crucial markers have arisen on several occasions in different genetic backgrounds. This is not unlikely if the markers are in rather unstable sequences, such as arrays of short tandem repeats. Again, inconsistencies would arise if the population under study was of mixed origin, recruited from originally separate groups. In that case, the population sample might exhibit a mixture of different gene associations, again with no reliable markers for QTLs.

In short, there are a number of reasons why the mapping of QTLs in outbred, for example human, populations is likely to be extremely difficult.

Conclusions and perspectives

The construction of detailed molecular maps of the genomes of various organisms, most notably that of humans, is making it increasingly possible to locate and clone the genes responsible for clear-cut heritable differences. Once a gene is cloned, only a moderate amount of work is needed to reveal sequence variations within it. And once a particular variation has been identified in one individual it becomes rather easy, through the use of an allele-specific oligonucleotide probe, to test for the presence of that same sequence difference in any other individual. This is obviously going to be relevant to people who, from their family histories, seem likely to be carrying particular known deleterious recessive alleles and at risk of having defective children.

Molecular screening of gene sequences in populations reveals a great deal of variation. Deletions, frameshifts and chain-termination codons within exons point clearly to functional deficiencies, but many other variants, such as synonymous codon replacements and most changes within introns, are almost certainly inconsequential. Amino acid replacements are more difficult to assess. Many proteins, some more than others, will tolerate a considerable amount of variation at certain positions in the polypeptide chains without any obvious effects on function. But it is extremely difficult to distinguish genuinely neutral changes from those of small effect, and whether there is an effect or not is likely to depend on the environment. 'Silent' variation in codons is so much more prevalent in populations than amino acid replacements that one must conclude that most of the latter reduce fitness.

Between the genes, especially in repetitive sequences, there is a great abundance of variation. Nearly all of this seems likely to be 'silent' phenotypically, though some no doubt affects gene regulation in various subtle ways, thereby contributing to quantitative variation in the phenotype. The clearly important role of all these DNA polymorphisms in quantitative genetics is to provide markers for the mapping of the quantitative trait loci, but a few of them could be QTLs in themselves.

Using molecular markers, some QTLs of larger effect can be mapped to particular chromosome regions, at least in organisms where inbred lines can be obtained (much less easily in humans). Even so, identification of QTLs in the DNA is still a formidable problem. In the first place their mapping will always be less accurate than that of genes that can be precisely scored, and even more megabases of DNA will need to be explored in order to find them. Secondly, the problem of recognition will be greater because of lack of clues as to gene function. Nevertheless, our example of the likely blood pressure gene in rat shows that good guesses can sometimes be made.

One objective of research on QTLs is to rescue them from that equivocal status and establish them as variants of regular genes, or perhaps regulatory regions, with defined sequences and functions. Once a QTL has been recognized in the DNA it should be possible, probing with an allele-specific oligonucleotide, to follow its transmission without ambiguity. This could, in favourable cases, add substantially to the efficiency of selective breeding of plants and animals. It is clear, however, that analysis of quantitative variation at the molecular level has to be selective, not comprehensive. There is no realistic prospect of obtaining a complete catalogue of DNA variants in any outbred population, still less of a complete description of their phenotypic effects.

References

Dryja, T.P., Hahn, L.B., Cowley, G.S., McGee, T.L. & Berson, E.L. (1991) Mutation spectrum of the rhodopsin gene among patients with autosomal dominant retinitis pigmentosa. *Proc Natl Acad Sci USA*, **88**, 9370–3.

Falconer, D.S. (1989) *An Introduction to Quantitative Genetics*, 3rd edn. Longman, London. (A valuable general reference.)

Higuchi, M., Antonarakis, S.E., Kasch, L. *et al.* (1991) Molecular characterization of mild-to-moderate haemophilia A: detection of the mutation in 25 of 29 patients by denaturing gradient gel electrophoresis. *Proc Natl Acad Sci USA*, **88**, 8307–11.

Hubby, J.L. & Lewontin, R.C. (1966) A molecular approach to the study of genetic heterozygosity in natural populations I. The number of alleles at different loci in *Drosophila pseudoobscura*. *Genetics*, **54**, 577–94.

Jacob, H.J.L., Lindpaintner, K., Lincoln, S.E. *et al.* (1991) Genetic mapping of a gene causing hypertension in the stroke-prone spontaneously hypertensive rat. *Cell*, **67**, 213–24.

Kreitman, M. (1983) Nucleotide polymorphism at the alcohol dehydrogenase locus of *Drosophila melanogaster*. *Nature*, **304**, 412–17.

Lander, E.S. & Botstein, D. (1989) Mapping Mendelian factors underlying quantitative traits using RFLP linkage maps. *Genetics*, **121**, 185–99.

Leckie, B. (1992) High blood pressure – hunting the genes. *BioEssays*, **14**, 37–41.

Ledbetter, D.H., Rich, D.C., O'Connell, P., Leppert, M. & Carey, J.C. (1989) Precise localization of NF1 to 17q11.2 by balanced translocation. *Am J Hum Genet*, **44**, 230–40.

Roberts, R.C., Bobrow, M. & Bentley, D.R. (1992) Point mutations in the dystrophin gene. *Proc Natl Acad Sci USA*, **89**, 2331–5.

Rommens, J.M., Ianuzzi, M.C., Kerem, B.-S. *et al.* (1989) Identification of the cystic fibrosis gene. Chromosome walking and jumping. *Science*, **245**, 1059–65.

Serre, J.L., Taillendier, A., Mornet, E. *et al.* (1991) Nearly 80% of cystic fibrosis heterozygotes and 64% of couples at risk may be detected through a unique screening for mutation by ASO reverse dot blot. *Genomics*, **11**, 1149–51.

Stuber, C.W., Lincoln, S.E., Wolff, D.W., Helentjaris, T. & Lander, E.S. (1992) Identification of genetic factors contributing to heterosis in a hybrid from two elite maize inbred lines using molecular markers. *Genetics*, **132**, 823–39.

Tsui, L.-C. (1992) The spectrum of cystic fibrosis mutations. *Trends Genet*, **8**, 392–8.

Zeng, Z.-B. (1992) Correcting the bias of Wright's estimates of the number of genes affecting a quantitative character: a further improved method. *Genetics*, **131**, 987–1001.

7

GENE INTERACTIONS AND THE GENETIC PROGRAMME

Introduction

Clues to gene function

As genome projects go to completion, it will become possible to identify all the genes of the selected organisms and deduce the one-dimensional structures of the proteins that they encode. The sequence information will often, though by no means always, reveal similarities to proteins with known functions, already in the computerized database. Furthermore, given the nucleotide and predicted amino acid sequences, oligonucleotide or antibody probes can be obtained for locating messenger RNAs (mRNA) and the products of their translation. In this way, as we see later in this chapter, the times and places of the activities of particular genes can be determined.

The structures and functions of the promoters that control specific patterns of gene expression can be investigated through the use of 'reporter' constructs, an example of which was introduced in Chapter 4 (see Fig. 4.15). Putative promoter sequences are joined to an open reading frame encoding an easily quantifiable enzyme, such as *Escherichia coli LacZ* (β-galactosidase) or *CAT* (chloramphenicol transacetylase), and introduced into the organism by some transformation procedure.

All of this can be done with the normal ('wild type') organism, without recourse to mutants. But to judge the role and importance of a particular gene it is necessary to see what happens when it is altered, either by random mutation or by controlled DNA manipulation. The successful analyses, to be described in this chapter, of gene function in *Saccharomyces cerevisiae*, *Drosophila melanogaster* and the flowering plant *Arabidopsis thaliana* were all based in the first instance on intensive mutant hunts, leading to the identification of genes that, when mutated, make a clear difference to the phenotype. After such genes have been identified and cloned, further information as to their functions can be obtained by controlled manipulation of their isolated DNA and its return to the organism. Possible 'engineered' mutations include both total knock-out of gene function (cf. Fig. 3.13) and enhancement of gene expression through increased copy number or attachment to a stronger promoter.

The maximum ambition of developmental genetics is to show exactly how the information in the genes is used to construct the entire organism – in other words, to elucidate the entire genetic programme. This involves far more than the description of individual gene functions. Genes never act in isolation; every aspect of the phenotype depends on a multitude of genes interacting in complex ways. It is with gene interaction that this chapter will be mainly concerned.

Molecular mechanisms of gene interaction

Interactions between different genes or gene products may be between DNA and protein, RNA and protein, or protein and protein.

The binding of proteins to the DNA of promoters and enhancers is one of the main keys to the understanding of gene regulation during development. Usually, several different proteins bind to the same regulatory element, and interact with each other as well as with the DNA. In Chapter 4 we reviewed some examples of complex promoters in the context of gene structure, and in this chapter we shall see their importance in development.

The genes encoding the RNA-binding proteins of the spliceosomes have universal functions in the splicing-out of introns. Some intron-splicing, however, is selective, making it possible to obtain different mature messengers from the same primary transcript. We saw one example in Chapter 4 (see Fig. 4.9), and others are constantly coming to light. Genes encoding RNA-binding proteins involved in selective intron-splicing have already been identified in *Drosophila* (see pp. 192–194).

Protein–protein interactions occur within a great variety of multi-protein complexes – ribosomes, spliceosomes, membrane structures, DNA and RNA polymerases, transcription and translation initiation complexes, and many others. Within each complex, the function of each component is more or less dependent on the others. Furthermore, many proteins are structurally and functionally modifed through the enzymic activities of other proteins. Phosphorylation and dephosphorylation, catalysed by protein kinases and phosphatases, are ubiquitous modes of control of protein function. The importance of acetylation, especially of histone proteins, is becoming increasingly apparent (see pp. 209–211).

Important clues as to the molecular mechanism of interaction between genes have often been obtained from their sequences. The DNA sequences may reveal open reading frames encoding proteins with DNA-binding motifs, or with affinities to known protein kinases, or other suggestive features.

Epistasis and its interpretation

The classical way of deducing connections between gene functions is through showing that the effect of a mutation in one gene is modified by mutation in another gene. *Epistasis* in the broad sense means that the double-mutant phenotype is not the simple sum of two single-mutant effects. In the extreme case in which the term is most often used, one mutation is *epistatic* in the sense that it more or less completely obscures the effect of the other – the double mutant resembles the epistatic single mutant.

There are several different general ways in which epistasis, in the strong sense, can work. One gene may provide the essential conditions for the activity of another, providing it with the metabolic intermediate on which its enzyme product acts, or with an activator essential for its transcription. In such cases, a loss of function of the first gene will make the second gene useless, and the single mutant will be just as disabled as the double mutant.

A second possibility is that the first gene inhibits the function of the second, perhaps by encoding a specific transcriptional repressor. In this situation, epistasis can work the other way round; a null mutation in the controlled gene will make the controller gene inconsequential, assuming (which will often not be true) there are no other genes under its control.

Any gene under the control of another, whether by activation or repression, can in principle mutate so as to work independently of control. Such liberating mutations will be more or less epistatic to mutations in the controlling gene, depending upon what else the latter has to do.

The interpretation of epistatic relationships has to be made in the light of the properties of the mutants concerned – whether they have decreased or increased gene product, and whether they are recessive or dominant. Null mutations in an activator gene decrease, and those in a repressor gene increase, the output of the gene under control. Null, or reduced-function mutations generally, are usually recessive. Mutations that free genes from control are usually dominant.

Examples from the yeasts

Control of the cell cycle – an example of a protein cascade

Most of our present information about the control of mitosis has been obtained from studies on yeasts, since mutants in which the process is modified are rather readily obtainable in these simple eukaryotes. Among temperature-sensitive mutants, able to grow at 25 °C but not at 35 °C, are many that are blocked at the higher temperature at specific stages of the cell cycle. In this brief account we will concentrate on the fission yeast *Schizosaccharomyces pombe*.

A gene that plays a central role in *S. pombe* mitosis is *CDC2*. Temperature-sensitive *cdc2* mutants may be blocked at higher temperatures either prior to DNA replication or between DNA replication and mitosis, depending on the mutant. The implication is that the wild-type *CDC2* protein is essential, in somewhat different ways, at both stages. The gene has been cloned and sequenced, and the predicted sequence of the protein product is typical of a group of protein kinases found in many organisms and called p34 on account of their molecular weight of about 34 000.

Two other genes have been identified whose protein products interact with the *S. pombe* p34. One is called *WEE1*, discovered and named in a Scottish laboratory; null mutations accelerate entry into mitosis so that the mutant cells are forced to divide when only part-grown. The amino acid sequence encoded by *WEE1* is again strongly suggestive of a protein kinase.

A second gene, *CDC25*, has the opposite effect of enhancing the activity of *CDC2*. It has been shown to encode a phosphatase that removes phosphoryl groups from *CDC2* protein. Loss of *CDC25* activity in temperature-sensitive *cdc25* mutants stalls the cell cycle between DNA synthesis

and mitosis – the cell keeps on elongating but does not divide. Over-expression of *CDC25* confers a wee phenotype even in the presence of wild-type *WEE1*, and such over-expression combined with a *wee1* mutation drives the cell into mitosis so prematurely that the effect is lethal.

The conclusion that *WEE1* and *CDC25* act through the *CDC2* p34 protein derives from the observation that certain amino acid-substituting mutations in *CDC2* bypass control by these genes. Two such mutations, *cdc2-2W* and *-3W*, confer a wee phenotype regardless of the state of *WEE1*. Presumably these amino acid replacements render p34 immune to inhibitory phosphorylation. Another *CDC2* mutation *cdc2-1W*, also giving a wee phenotype, is epistatic to loss of function of *CDC25*; presumably the amino acid substitution here makes p34 active even without dephosphorylation by the *CDC25* enzyme.

CDC25 activity appears to be coupled in some way to DNA replication. The *cdc2-1W* mutant, in which the *CDC2* p34 protein has been freed from dependence on *CDC25*, proceeds into mitosis even when DNA synthesis has been blocked by the inhibitor hydroxyurea. Under these circumstances the cells attempt to undergo mitosis, with formation of division spindles and so on, while they still have only a single complement of DNA. In *cdc2-2W* and *-3W*, which still require *CDC25*, inhibition of DNA synthesis blocks mitosis, as it does in the wild type. It seems that *CDC25* or its protein product, normally necessary for *CDC2* function and hence for mitosis, is only activated once DNA synthesis has been completed.

The *WEE1* protein product, itself a protein kinase, is negatively controlled by another protein kinase-encoding gene, *NIM1* (New Inducer of Mitosis). Over-expression of *NIM1* results in a wee phenotype, and its loss of function in a temperature-sensitive mutant blocks mitosis. Neither increase nor loss of function of *NIM1* has any effect in the absence of a functional *WEE1* allele. The scenario is that the *NIM1* kinase phosphorylates and inactivates the *WEE1* kinase, which phosphorylates and inactivates the *CDC2* kinase (Wu & Russell, 1993).

Note that in the system of mitotic control, as analysed so far (Fig. 7.1), there are no effects on gene transcription. The relevant genes are

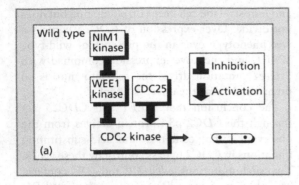

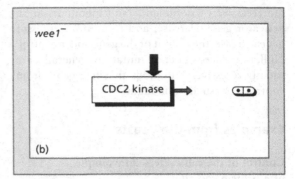

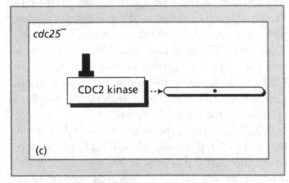

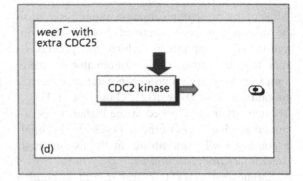

Fig. 7.1 Interactions between gene products in the control of cell division in *Schizosaccharomyces pombe*, and the effects of some mutations. (a) Wild type. The NIM1 kinase phosphorylates and inactivates the WEE1 kinase which phosphorylates and inactivates the CDC2 kinase, which provide a signal for cell division. The CDC25 phosphatase dephosphorylates and activates CDC2 kinase. The joint effect of these controls is to activate cell division after a period of growth sufficient to produce daughter cells of normal size (sketched to the right). (b) In a null *wee1* mutant CDC2 kinase is hyperactive and cell division occurs before cells have grown to normal size. (c) In a temperature-sensitive *cdc25* mutant, at the restrictive temperature, CDC2 kinase is not activated, and the cell grows indefinitely without division. (d) Controlled overproduction of CDC25 phosphatase in a *wee1* mutant is lethal, since division occurs almost without growth. Adapted from Forsberg & Nurse (1991).

interacting through their protein products – by proxy, as it were. This is in marked contrast to the extensively studied systems of control of metabolism. We consider two such systems, both in *Saccharomyces*, in the following sections.

Galactose metabolism in budding yeast – transcriptional regulation and protein–protein interaction

Saccharomyces cerevisiae seems a very simple organism from the morphological point of view, but it has great versatility when it comes to the utilization of nutrients. Like all other opportunistic micro-organisms, it produces from its repertoire of enzymes only those that are immediately necessary for growth. For example, when called upon to use the sugar galactose as source of carbon it produces a set of enzymes that can convert galactose to the more common sugar glucose. When glucose itself is available in the growth medium, the enzymes for galactose utilization are hardly produced at all.

Three enzymes are necessary for the galactose-to-glucose conversion: galactokinase, galactose-1-phosphate UDP transferase, and UDP-galactose epimerase, acting in that order. The three enzymes are encoded respectively by the genes *GAL1*, *GAL7* and *GAL10*, which map in a tight cluster but are separately transcribed. An unlinked gene,

---

Box 7.1 Some features of DNA-binding proteins

. . . .

Class	Encoding genes (examples)	Characteristic motif	Role in DNA binding
Homeobox	*Drosophila Antp, Ubx, bcd, ftz, en*, Mouse *Hox* genes	*c.* 40-amino acid sequence forming helix-turn-helix	One helix slots into major groove of DNA duplex
Zinc-finger	Yeast *GAL4*, *Drosophila hb*, *Kr, kni*	One or several sequences with histidines and cysteines placed so as to bind Zn, e.g.,	Pins the protein in configuration suitable for DNA binding
Leucine zipper	Yeast *GCN4*	*c.* 50 residues with leucines at 7-residue spacing	Association of two helices by their water-repellent surfaces – stabilizes helical structure for DNA binding

GAL2, encodes a membrane-bound protein, galactose permease, responsible for uptake of galactose into the cell. Transcription of all four genes is induced by galactose in the growth medium and repressed by glucose. Mutations in any of them can block utilization of galactose.

Another class of galactose non-utilizing mutants, *GAL4*, is deficient in ability to transcribe all four of the above genes. *GAL4* has been cloned and sequenced, and the protein predicted from its open reading frame has a typical DNA-binding domain of the 'zinc-finger' class (Box 7.1). The GAL4 protein binds specifically to a DNA sequence motif that is common to the upstream regions of *GAL1, 2, 7* and *10*. Activation of transcription is due to another GAL4 protein domain, notable for its high proportion of acidic amino acid residues. This acidic domain acts in concert with a number of other more general transcription factors, and is, in itself, rather unspecific in its action. Its targeting on *GAL* genes is due to the specific DNA-binding sequence to which it is joined. If, by *in vitro* manipulation of the encoding DNA, the DNA-binding motif of the GAL4 protein is replaced by that of an activator of another class of genes, then the GAL4 acidic domain will activate those other genes and not *GAL1*, *2*, *7* and *10*.

Why then does *GAL4* only work in the presence of galactose? The first clue to the answer was the finding of recessive mutations, attributed to a new gene *GAL80*, that had the effect of making the transcription of *GAL1/2/7/10* constitutive – that is, constant irrespective of carbon source. The recessivity of these mutations implied loss of function, and that function appeared to be the repression of transcription of *GAL1/2/7/10*. The second clue was the discovery of a second, much rarer, class of dominant constitutive mutants with amino acid replacements in a certain region within the activation domain of the *GAL4* protein. The interpretation, now strengthened by more direct protein analysis, is that the *GAL80* protein binds to the *GAL4* protein so as to block its activation function. The amino acid replacements in the *GAL4* dominant constitutive mutants disrupt the region to which the *GAL80* protein specifically binds.

This still leaves unanswered the question of how the system responds to galactose. The likely answer emerged from study of another gene, *GAL3*. The phenotype by which *gal3* mutants were identified was very slow adaptation to growth on galactose. The most surprising property of this delayed response was its discontinuous, mutation-like nature. When, after a long lag, a cell did start using galactose efficiently, it formed a rapidly growing clone of cells sharing this ability, even while the majority of cells on the galactose-containing plate were still stuck in the lag phase. This phenomenon, when first studied as long ago as the late 1940s, prompted some adventurous but now discounted speculations about genes in the cytoplasm. A recent return to the study of *GAL3*, but now at the molecular level, puts this interesting

epigenetic effect (see p. 205) on a firmer theoretical basis.

The cloning and sequencing of *GAL3* led to some surprises. The first was the high degree of sequence similarity that the predicted *GAL3* protein had to galactokinase (the product of *GAL1*), even though it had no galactokinase activity. This suggested that *GAL3* protein might bind galactose. Secondly, either *GAL1* or *GAL3*, under the control of a strong promoter in a multicopy plasmid, would derepress *GAL7* transcription in the absence of galactose, though even better in its presence. These observations have led to the hypothesis that galactokinase, in addition to its enzymic function, can duplicate the function of *GAL3* protein, which is now thought to be to bind to *GAL80* protein to counter the latter's inhibitory effect on *GAL4*. It is postulated that galactokinase and *GAL3* protein each exist as an equilibrium mixture of two interconvertible forms; one able to counter the effect of *GAL80* protein and the other not. Binding of galactose may shift the equilibrium to the first form, with consequent release of GAL4-activated transcription. On this hypothesis, when either *GAL1* or *GAL3* is over-expressed there is enough anti-GAL80 protein in the cell to be effective, even in the absence of galactose (Fig. 7.2).

This hypothesis can explain the lag in galactose utilization by *gal3* mutants, and also the heritability, through a cell lineage, of the ability to use galactose once acquired. In the absence of *GAL3* protein, derepression of the GAL system depends on the *GAL1* product, galactokinase. But until derepression is under way there will be hardly any galactokinase to do the job, since *GAL1* will be repressed. With only a few galactokinase molecules per cell, there will be significant statistical fluctuation in number from one cell to another. In cells that, by chance, get above a certain threshold, galactokinase will begin to derepress the transcription of its own gene, and the whole GAL system will then go into a phase of exponential build-up leading to a self-sustaining steady state – an example of a *positive feedback loop*.

All this promises to explain the induction of the system of galactose utilization by galactose. Its repression by glucose, an aspect of control not included in Fig. 7.2, involves another gene *MIG1*,

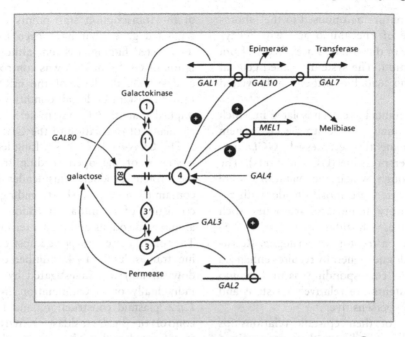

Fig. 7.2 Control of galactose metabolism in *Saccharomyces cerevisiae*. GAL4 protein activates (⊕) transcription by binding to the promoters (small circles) of *GAL1, 10* and *7*, and *MEL1*, the latter encoding melibiase, an enzyme needed for the utilization of a galactose-containing disaccharide. GAL80 protein (80) binds to GAL4 protein (4) to inhibit its function. The GAL3 protein (3), and also the GAL1 protein (1, galactokinase), bind to galactose (gal) and supposedly undergo conformational changes to forms 3′ and 1′ which block the GAL4–GAL80 interaction and hence free GAL4 protein from GAL80 inhibition. Note that there are two potentially self-sustaining feedback loops. *GAL4* activates *GAL2* to produce a permease that brings galactose into the cell to permit the expressed GAL3 protein to release GAL4 from GAL80 inhibition. In default of *GAL3* function, GAL4 protein activates *GAL1* to produce galactokinase which, in presence of galactose, provides an alternative way of cancelling the negative effect of GAL80 on GAL4. This second loop probably takes longer to become established, accounting for the delayed switch from no growth to growth on galactose seen in *GAL3* mutants. Data from Johnston (1987) and Bhat & Hopper (1992).

which encodes a DNA-binding protein of the 'zinc-finger' class. The *MIG1* protein represses transcription of *GAL4*, *GAL1* and probably the other genes required for galactose utilization. Null mutations of *MIG1* cause substantial (though not complete) release of the GAL system from glucose repression. The next step upstream, the connection between the MIG1 protein and glucose, remains to be worked out.

The GAL system of yeast illustrates three different levels of gene control: control of transcription (both positive and negative), control via protein–protein interactions, and maintenance of gene activity by positive feedback. Our next example brings in a fourth level – control of mRNA translation.

General control of amino acid biosynthesis – a protein kinase cascade that regulates translation

Saccharomyces cerevisiae, like most other fungi, can synthesize all the amino acids required for protein synthesis when it needs to do so, but it uses ready-made amino acids so far as it can. When growing in an environment rich in amino acids, it makes only minimal levels of the numerous enzymes required for amino acid biosynthesis, but in response to amino acid starvation the levels of most of these enzymes are increased (derepressed) by factors ranging from twofold to a 100-fold. Deprivation with respect to any one of a number of different amino acids evokes an increase in the enzymes required for synthesis of them all.

Presumably, in nature as opposed to the laboratory, a shortage of one amino acid will always imply a shortage of the others, since all come from protein degradation. The system is called general control of amino acid biosynthesis, abbreviated to GC.

Numerous mutants have been isolated in which the enzymes normally subject to the GC system are either permanently derepressed (*GCD* mutants) or non-derepressible (*GCN* mutants). The phenotypes through which the mutations were originally recognized are mostly to do with response to inhibitory amino acid analogues, such as triazole-alanine, a histidine analogue, and 5-fluorotryptophan, a tryptophan analogue. Resistance to such analogues generally requires enhanced synthesis of the corresponding natural amino acids; *GCD* mutants are relatively resistant and *GCN* mutants hypersensitive.

On the basis of their epistatic relationships the genes can be sorted into three groups in a hierarchy of control: first *GCN2* and *GCN3*, then several *GCD* genes, and finally *GCN4* (Fig. 7.3a). The wild-type *GCN4* gene is unconditionally necessary for derepression and the non-derepressible phenotype of a null *gcn4* mutant is unaffected by any mutation in any other *GCN* or *GCD* gene. The *GCD* genes are evidently jointly involved in blocking the activity of *GCN4* on rich growth medium, and their mutations make no difference if *GCN4* is already inactivated by mutation; *gcn4* are epistatic to *gcd* mutations. The latter are in turn epistatic to *gcn2* and *gcn3*; double mutants *gcd gcn2* or *gcd gcn3* are derepressed. One can interpret this as meaning that the wild-type *GCN2* and *GCN3* genes act to counteract the negative effect on *GCN4* of the wild-type *GCD* genes, an effect that in *gcd* mutants has in any case been lost (Fig. 7.3c). That is as far as it was possible to go without molecular studies. Cloning and sequencing of some of the genes led to a much more detailed scenario.

Not unexpectedly, the *GCN4* protein turns out to be a transcriptional activator, comparable in most ways to *GAL4* protein, though with a different kind of DNA-binding domain – 'leucine zipper' (Box 7.1). The DNA sequence to which it binds is found, with minor variations, upstream of the transcription start-points of all the GC-regulated genes that have been looked at. The unexpected finding was that although the translation of *GCN4* mRNA was controlled by amino acid supply, the level of the mRNA itself was rather constant under all conditions. This clearly implied that the GC system acted to control the efficiency of translation of the *GCN4* mRNA.

The *GCN4* mRNA has a long leader sequence upstream of the open reading frame encoding the protein product. This leader is unusual in containing four very short reading frames, each consisting of an initiation codon, two or three amino acid codons and then a termination codon. The role of these unexpected upstream open reading frames (uORF1–4, numbered upstream-to-downstream) was investigated by excising them individually or in combination from a *GCN4/LacZ* plasmid construction, and looking at the control of β-galactosidase activity in plasmid-transformed cells. The results show that the presence of one or other of the two most downstream uORFs is necessary for repression of translation by amino acids; when they are both excised, translation proceeds at an enhanced level whether amino acids are present in the growth medium or not. Loss of uORF1, while it hardly changes the low level of mRNA translation in the presence of amino acids, strikingly reduces the positive response to amino acid deprivation. Loss of uORF2 has no dramatic effect. These results seem to mean that ribosomes bind at the most upstream initiation codon in uORF1, but then, in order to gain access to the *GCN4* reading frame, need some special help to bypass uORFs 3 and 4.

In trying to understand what might be happening here, it is necessary to draw on the results of extensive biochemical studies. Briefly, translation initiation in eukaryotes involves a protein complex, eIF2, which binds the methionyl-tRNA which provides the methionine for the N-terminus of the polypetide chain. In its active form, eIF2 is also bound to guanosine triphosphate (GTP), which, as initiation occurs, is hydrolysed to the much less energy-rich diphosphate (GDP). To function again, eIF2 has to be recharged with GTP, a reaction catalysed by an accessory protein complex eIF2B. The recharging of eIF2 is inhibited

Fig. 7.3 (a) Epistatic relationships among genes involved in regulation of general amino acid biosynthesis. Mutants in *GCN4* are unable to derepress biosynthesis whatever the status of the other genes. Mutants in *GCD* genes are constitutively derepressed (given *GCN4* function) regardless of *GCN2* and *GCN3*. *GCN2* and *GCN3* mutations result in inability to derepress, but their effects are over-ridden by *GCD* mutations. *GCN2, 3* are considered to inhibit the action of the *GCD* genes, which in turn repress *GCN4*. (b,c) Molecular hypothesis for general control of amino acid biosynthesis in *Saccharomyces cerevisiae*. (b) Ribosomes initiate and then terminate translation at an upstream short open reading frame (ORF) of the *GCN4* mRNA leader sequence, and then track down the leader sequence 'looking' for another initiation codon. Initiation requires guanosine triphosphate (GTP) bound to the ribosome-associated initiation complex eIF2, and involves its hydrolysis to guanosine diphosphate (GDP). The accessory protein complex eIF2B replenishes the GTP in time for the ribosomes to reinitiate at the downstream short ORFs. But they are then unable to reinitiate efficiently at the *GCN4* initiation codon. (c) When amino acids are scarce, accumulated free transfer RNA binds to and activates the GCN2 protein kinase, which phosphorylates a component of eIF2-GDP so that it can no longer be acted upon by eIF2B. As a result, GTP is not quickly replaced, the ribosomes do not reinitiate efficiently at the downstream leader ORFs, but are better able to do so at the *GCN4* ORF. *GCD1, 2, 6, 7* and also *GCN3* all encode components of eIF2B; GCN3 protein is thought somehow to mediate the inactivating effect of GCN2 kinase. Based on Hinnesbuch (1988, 1993), Dever *et al.* (1992) and Cigan *et al.* (1993).

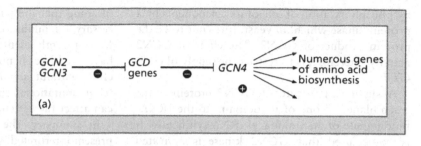

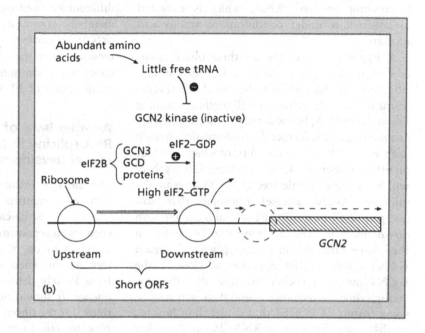

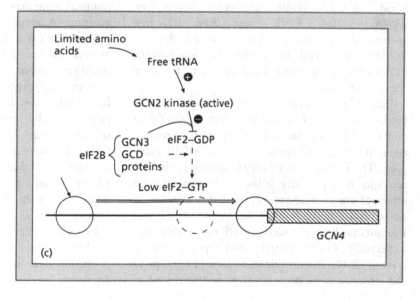

by phosphorylation of one of its components by a protein kinase which, in yeast, turns out to be the protein product of *GCN2*. The effect of *GCN2* activity, therefore, is to restrict the supply of eIF2-GTP.

A significant feature of *GCN2* protein is the resemblance of one of its domains to the tRNA-binding site of an aminoacyl-tRNA synthetase. It is conjectured that *GCN2* kinase is activated by binding to free tRNA, which is expected to accumulate under conditions of amino acid starvation.

A hypothesis that fits all these observations is shown in Fig. 7.3b and c. The idea is that ribosomes discharged from the end of one reading frame in the leader sequence will continue scanning down the mRNA, but will only initiate translation at a reading frame further downstream if recharged with eIf2-GTP, the availability of which depends on eIf2B. If there is plenty of active eIf2B (as there will be if there is little free tRNA to activate the inhibitory *GCN2* kinase), ribosomes that have just passed through uORF1 will also engage with downstream uORFs including uORF4. But then they have difficulty in engaging with the main *GCN4* ORF; uORF4 may be too close to the *GCN4* initiation codon to allow the ribosome enough time for recharging, and there may also be some feature of uORF4 that causes the ribosome to dissociate from the mRNA. If, on the other hand, eIf2-GTP is limited (amino acids scarce, free tRNA plentiful, *GCN2* kinase active, eIf2B less active), the ribosomes, having passed through uORF1, will tend to ignore the downstream uORFs but still get recharged in time to translate the *GCN4* ORF.

Recent biochemical findings (Cigan *et al.*, 1993) help explain the effects of mutations in the *GCD* genes and *GCN3*. Several *GCD*-encoded proteins turn out to be components of the eIF2B complex. The impairment of eIF2B activity in the gcd mutants is apparently sufficient to simulate the effect of amino acid starvation, but not so severe as to make the cell inviable; in fact, some *gcd* mutants do have reduced growth rates. More unexpectedly, *GCN3* protein also appears to be a part of eIf2B. The non-derepressible phenotype of the original *gcn3* mutants is now interpreted as

meaning that the wild-type *GCN3* protein is necessary for inhibition of eIf2B by *GCN2* kinase. More recently identified mutations, called *gcn3^c*, have a gcd-like phenotype, and appear to inactivate eIf2B without any help from *GCN2*. Thus different mutational changes in the *GCN3* protein can affect the system in opposite ways.

In summary, the GC system, at least as at present interpreted, shows a combination of three different kinds of signal, acting successively: (i) a metabolic signal, in this case high levels of free tRNA; (ii) protein phosphorylation, conveying a message to the translation system; and (iii) the concerted activation by *GCN4* protein of the transcription of 30 or more separate genes.

Another level of control – RNA splicing in *Drosophila* sexual development

The ends of introns are marked by 5' and 3' terminal consensus sequences – the donor and acceptor sites, so-called in reference to the biochemical mechanism. The acceptor site consensus amounts to no more than PyAG at the intron terminus preceded, some tens of bases upstream, by a loosely defined branchpoint or 'lariat' sequence. It should be not too difficult, one might think, for the ribonucleoprotein spliceosome complex to fail to recognize one of these rather minimal signposts and to fasten instead on the acceptor sequence of the next intron downstream, splicing out two introns and the intervening exon as well. Through errors of this kind one gene could yield two or more alternative RNAs encoding different polypeptide chains, perhaps with different functions. It would be surprising if living organisms did not sometimes use controlled splicing 'errors' as a means of modulating gene function. There are, in fact, an increasing number of examples of this kind of control (we saw one in Chapter 4, see Fig. 4.9), and it plays a particularly important part in sexual differentiation in *Drosophila*.

In flies, gender is determined by the ratio of X-chromosomes to sets of autosomes, with the Y-chromosome just carrying a few genes necessary for sperm function. The switching of development

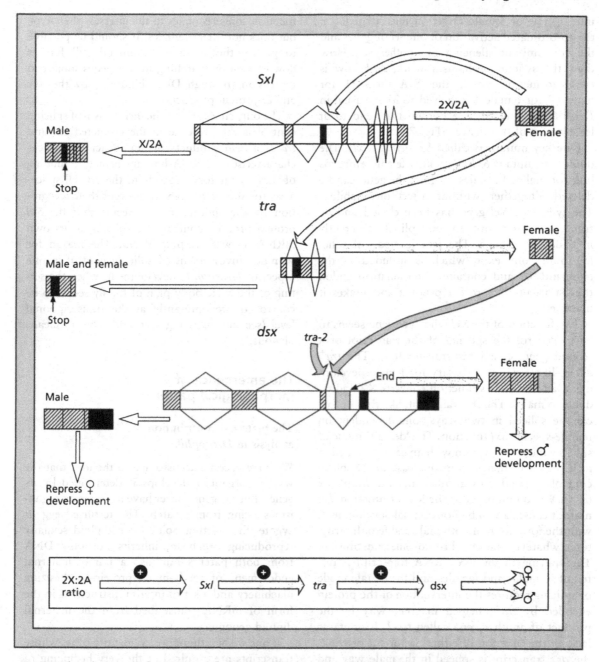

Fig. 7.4 The splicing cascade in *Drosophila* sexual development. The mode of splicing of the *Sxl* transcript is initially directed, in ways not yet understood, by the X/autosome ratio. The male mode of splicing generates a messenger RNA (mRNA) with an early stop codon; the female mode generates a mRNA encoding a protein that both maintains the female mode of splicing of *Sxl* RNA and directs female-specific splicing of the *tra* transcript. The female *tra* protein product, together with the *tra-2* product, determines female-specific splicing of the *dsx* transcript. The two alternative *dsx* mRNAs are both functional, in male and female development respectively. *End* indicates a termination sequence in the female-specific product. The heavy arrows show the female-specific positively-driven cascade. The light arrows indicate the default (male) modes of splicing. After Baker (1989).

into a male or female direction must depend on the co-ordinated activation of one set of genes and the concomitant silencing of another set. How does the switch mechanism work, and how is the information about the X/A ratio (1 for males, 2 for females) conveyed to it? As usual in *Drosophila*, mutants are known that lead to at least a part of the answer (Fig. 7.4).

One key mutant is called *Sex-lethal* (*Sxl*). The dominant mutant *Sxl* allele kills female embryos but not males. In males, in fact, the gene can be deleted altogether without affect on viability. The wild-type *Sxl* gene has been cloned and its transcripts turn out to be spliced differently in male and female. The female mode of splicing omits one exon which is included in the male mRNA and contains a termination codon that shortens the protein product and makes it inactive.

The function of the *Sxl* wild-type gene seems to be to control the splicing of the transcript of a second gene, called *transformer* (*tra*). The *tra*⁺ allele, like *Sxl*⁺, is necessary for female development, and loss of its function converts XX individuals to males. The *tra* transcript, like that of *Sxl*, can be spliced in two ways, one dependent on female-specific *Sxl* function. The 'default' mode of splicing includes an exon with an early stop codon in it. Yet a further step in this cascade of splice-control is revealed by another mutant, *doublesex* (*dsx*). When homozygous the loss-of-function *dsx* mutant causes an ambiguous sexual development, with the formation of both male and female structures whatever the sex chromosome constitution. The splicing of the *dsx*⁺ RNA transcript is different in males and females and the female mode of splicing requires the intervention of the protein encoded by *tra*⁺, helped in some way by the product of another gene called *tra-2*. In mutants lacking the function of either *Sxl* or *tra* or *tra-2*, the *dsx* transcript is spliced in the male way and male development ensues regardless of the sex chromosomes. The bisexual phenotype of *dsx* mutants implies that both the male and the female *dsx*⁺ mRNAs yield functional proteins, the male protein repressing female and the female protein repressing male development. Another gene, *intersex* (*ix*), apparently participates with *dsx* in the repression of male differentiation. Loss of *ix*

function converts males to the intersex phenotype but does not affect females. It should be possible to pursue the chain of command still further downstream by searching for the genes subject to repression (through DNA binding?) by the *dsx* and *ix* protein products.

The only clue so far to the mechanism of splicing control is the presence in the predicted *Sxl* and *tra-2* protein products of amino acid sequences characteristic of RNA-binding proteins. Binding of these regulatory proteins to the crucial splice-acceptor sites is thought to prevent their recognition by the spliceosome. It seems that the *Sxl* protein product controls the splicing of its own mRNA as well as that of *tra*. The reason for so much involvement of splicing control in one aspect of *Drosophila* development, when the setting of the basic body plan of the fly seems to be controlled predominantly at the transcriptional level (see the following section), is by no means obvious.

The emergence of morphological pattern

The maternal contribution – analysis in *Drosophila*

We have become accustomed to the idea that the way an organism develops is determined by its genes. But the genes never have to create an organism starting from scratch. The fertilized egg or zygote, the starting point for a diploid sexually reproducing organism, inherits not only DNA from both parents but also a purely maternal endowment of ready-made protein-synthesizing machinery and a set of interim instructions in the form of mRNA transcribed from the maternal diploid genome.

If one wishes, therefore, to identify genes whose transcripts are required at the very beginning of development, it is a good strategy to look for mutant phenotypes that are expressed in the early embryo but are determined by the maternal rather than the zygotic genotype. Following intensive hunts, a number of such mutants have been discovered in *D. melanogaster*, which is currently the prime organism for investigations of gene action in development. The genes concerned are called

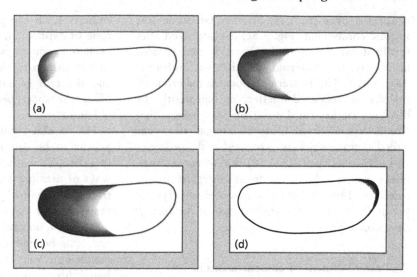

Fig. 7.5 Domains of expression of some genes in the *Drosophila* embryo: (a) messenger RNA (mRNA) of the maternal-effect gene *bicoid* detected by a DNA probe; (b) protein product of *bicoid*, detected by an antibody probe; (c) zygotic mRNA of the 'gap' gene *hunchback* in the absence of the normal maternal contribution to *hb* mRNA; (d) mRNA of the maternal-effect gene *nanos*. In each case the anterior end of the embryo is to the left and the posterior end to the right. Drawn from photographs from St. Johnston & Nusslein-Volhard (1992).

maternal effect genes. They are recognized by recessive mutations which, when homozygous in a female, make all the progeny inviable whatever the genotype of the male parent.

A particularly well-studied example is *bicoid* (*bcd*). If this gene is not functional in the mother, all embryos are grossly deficient in those structures developing from the anterior end of the egg; only posterior structures are recognizable. The gene has been cloned and sequenced, as have all the *Drosophila* genes to be mentioned in this chapter. Its predicted protein product includes a 60-residue sequence conforming to a consensus called a *homeobox*, first found in genes, such as those of the bithorax complex (see p. 202), deficiencies in which have homeotic (or homoeotic) effects – that is transformations of one set of body structures into another. The homeobox sequence forms the DNA-binding domain of a class of proteins concerned in transcriptional control (see Box 7.1). When *bcd* mRNA is labelled in the early embryo with a complementary single-stranded DNA probe, it is seen to be concentrated at the anterior end and virtually absent from the posterior end (Fig. 7.5a). The *bicoid*-encoded protein, detected by means of a labelled antibody, shows a steep anterior-to-posterior gradient of decreasing concentration (Fig. 7.5b).

The *bcd* protein product has a role in helping to regulate transcription of a number of other genes.

One that seems to be of particular importance is *hunchback* (*hb*), recessive lethal mutations in which confer an anterior-deficient embryonic phenotype rather similar to that due to absence of maternal *bicoid* activity. The *bicoid* function is epistatic to *hunchback*; in *bcd*-negative embryos *hb* mRNA is not present at all. Several different observations support the view that the *bicoid* protein is a transcriptional activator of *hb*. First, when *bcd* function is missing, neither supplied maternally nor through the male sperm nucleus, no detectable *hb* mRNA is formed. Secondly, when the *hb* promoter region, a sequence of a few hundred base-pairs upstream of the transcription start-point, is joined to *lacZ*, and the construct introduced into the *Drosophila* genome using the P-element as a vector for transformation (see p. 122), embryos inheriting the hybrid gene show b-galactosidase activity that is strong at the anterior end but decreases steeply towards the posterior – a pattern mirroring that of *bicoid* protein (Fig. 7.5c). Thirdly, 'footprinting' experiments show that *bicoid* protein, made in *E. coli* by a *bcd*[+] expression plasmid (see p. 75), binds to several sites within the *hb* promoter.

The establishment of the anterior-to-posterior decreasing gradient of *hb* protein is actually a more complicated matter than the last paragraph would suggest. There are, in fact, two sources of *hb* mRNA in the early embryo – much of it is carried over from the mother but some (the zygotic component, Fig. 7.5c) is synthesized after fertilization. Whereas the latter is concentrated anteriorly, the maternal *hb* mRNA is uniformly distributed. The posteriorly located maternal *hb* mRNA is rendered ineffective (apparently by blocking of its translation) by some mechanism that is dependent on another maternal-effect gene called *nanos*, which is expressed at the posterior end of the embryo (Fig. 7.5d). Still other maternal-effect genes, which we must disregard in the interests of brevity, are also involved, in ways not yet understood, in the stabilization of the *bicoid/hunchback* gradient.

The establishment of compartments in *Drosophila* development

The next steps in the anterior-posterior patterning of the *Drosophila* embryo are largely under the control of the so-called *gap* genes, which include *hunchback* (*hb*), *giant* (*gt*), *Krüppel* (*Kr*), *knirps* (*kni*) and *tailless* (*tll*). All of these encode proteins with sequences indicative of DNA binding – not homeobox sequences but so-called 'zinc-finger' or (in the case of *gt*) the leucine-zipper motif (Box 7.1). With the exception of *hb*, the gap genes are expressed only zygotically – that is after fertilization. They were recognized through their recessive lethal mutations, the effects of which, when homozygous, are to create large gaps in the normal sequence of structures along the anterior–posterior axis of the embryo. They are transcribed in the early embryo up to and including the blastoderm stage, when the original zygote nucleus has undergone multiple divisions to form a continuous single layer of cells at the surface of the embryo. Each gap gene has its own well-defined zone of expression, detected either as mRNA or protein product (Fig. 7.6a). Attention is particularly drawn to the centrally located *Kr* band and the two *gt* bands of expression, with *kni* filling the space between *Kr* and the posterior *gt* band.

The localized *hunchback* and *bicoid* proteins appear to be determinants of the zones of expression of the other gap genes. There are several ways of manipulating the concentration and distribution of *hunchback* protein. It can be supplied maternally or zygotically or both. The maternal component, normally concentrated at the anterior end, can be uniformly distributed by removing *nanos* function. The level of maternally-contributed hb protein is increased if the maternal parent is endowed with additional copies of *hb*+ by P-mediated transformation (see p. 122). Each of these manipulations affects the distributions of the gap gene protein products (Fig. 7.6b). Overall, the results are consistent with a model in which each gap gene is expressed within a certain range of *hb* (or *hb* and *bcd*) protein concentrations. Too low down the *bcd/hb* gradient they are not activated, and too high up they are repressed; the critical concentrations defining the anterior and posterior boundaries are different for each gap gene.

To strengthen this model it is necessary to find out what gene products do in fact bind to gap gene operator regions to activate or repress transcription. Such investigations tend to use fusions of the operators, or selected segments of them, to a suitable 'reporter' gene (e.g., *LacZ*), introduced into the genome by P-mediated transformation. It has been shown, for example, that the transcrip-

Fig. 7.6 (*opposite*) (a) Distribution of gap gene proteins along the long axis of the *Drosophila* embryo at the blastoderm stage. Boundaries are indicated between potential segments; three head segments, three thoracic segments (T1–3) and eight abdominal segments (A1–8). The vertical stippled stripes on the diagram above and on the sketch of the embryo below show expression of the pair-rule genes *even-skipped* (*eve*) and *hairy* (*h*), respectively; *h* is expressed in stripes that lie slightly anterior to those of *eve*. Adapted from Pankratz & Jäckle (1990).

(b) Changes in concentrations and distributions of *Krüppel* (*Kr*) protein in response to changes in *hunchback* (*hb*) protein distribution brought about as indicated to the left of the diagrams. (i) Wild type; *Kr* expressed in a central band where the *hb* protein level is neither too low nor too high. (ii) The maternally contributed *hb* protein is, in the absence of *nanos* function,

(*continued*)

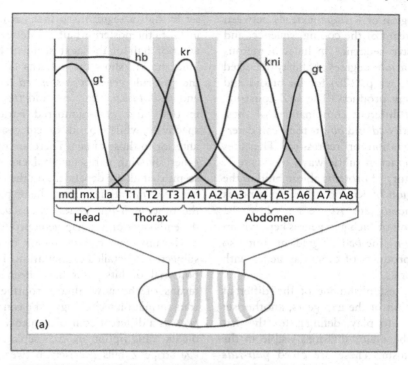

(a)

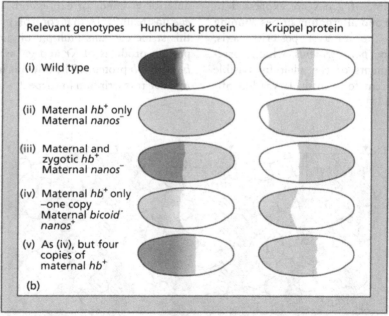

(b)

distributed throughout the embryo at a level appropriate for *Kr* expression – but at the anterior tip *Kr* is repressed by the *bicoid* product. (iii) The addition of zygotic *hb* product makes the total *hb* protein level too high for *Kr* expression in the anterior half of the embryo. (iv) A low level of *hb* protein of maternal origin is present at a suitable level for *Kr* expression in the anterior third of the embryo. The *hb* protein, and consequently the *Kr* protein, are absent from the posterior two-thirds because of the action of *nanos* (see text). (v) Boosting the level of *hb* protein by additional maternal gene copies pushes the zone of expression of *Kr* further towards the posterior. The single-dose *hb*⁺ in (iv) was because of maternal heterozygosity, and the extra copies in (v) were introduced by P-mediated transformation (see p. 122). Diagrams redrawn from Lawrence (1992).

tional control region of *Krüppel* extends between 1 and 4 kb upstream of the coding sequence, and that this extensive sequence includes numerous binding sites of various degrees of affinity (detected by 'footprinting', see p. 120) for the bicoid and hunchback protein products. The *lacZ* constructions show that different combinations of sites, in interaction with *bcd* and *hb* proteins, can determine either activation or repression. The sites undoubtedly interact in subtle ways.

The full inventory of proteins able to bind to the long control region of *Kr* and other gap genes is probably still incomplete. As we see below, the band of expression of each gap gene is kept within bounds not only by the *bcd/hb* gradients but also by the protein products of other gap genes with adjoining territories.

Following the establishment of the different zones of expression of the gap genes, another set of genes comes into play, defining, for the first time, some of the major divisions visible in the developed organism. These are called *pair-rule* genes, because pairs of genes are expressed in mutually exclusive and alternating stripes running perpendicular to the anterior–posterior axis. These stripes define *parasegments*, which correspond to the segments of the adult fly (visible in somewhat different form in the larva) but off-set by half a segment, so that each parasegment includes the posterior half of one segment and the anterior half of the next segment back. Figures 7.6a and 7.7 show the patterns of expression of the pair-rule genes *even-skipped* (*eve*), *hairy* (*h*) and *fushi-tarazu* (*ftz*); *eve* and *ftz* are expressed in odd- and even-numbered parasegments respectively, while *h* bands of expression lie slightly anterior to those of *eve*. There are other pair-rule genes showing parasegmental expression, but it seems that all are dependent on the prior action of *eve*; *eve* mutations affect the expression of all the other pair-rule genes while *eve*[+] expression depends only on the gap gene products.

The upstream control region of *eve* has been subjected to detailed examination. The regulatory potential of bits of it have been evaluated by means of the now almost routine *lacZ*-fusion/transformation technology. The conclusion is that *eve* has a different control sequence for switching on its transcription in each separate stripe. The *eve* stripe 2 falls precisely between the *Kr* and anterior *gt* bands of expression. The segment of *eve* upstream DNA that controls *eve* expression in this stripe contains multiple binding sites for the protein products of *Kr* and *gt*, as well as for the *bcd* and *hb* proteins. The obvious interpretation is that *eve* transcription in stripe 2 is activated by the

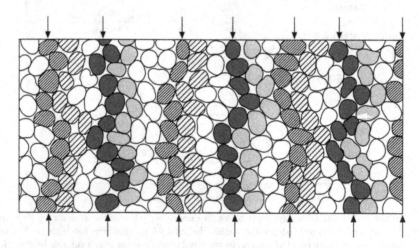

Fig. 7.7 The expression of the pair-rule genes *even-skipped* (*eve*) and *fushi-tarazu* (*ftz*) in alternating parasegments of the *Drosophila* embryo at the blastoderm stage. The *eve* and *ftz* protein products were detected by specific antibodies fluorescent in different colours (represented by hatching and stippling), and appear in odd-numbered and even-numbered parasegments respectively. The arrows indicate the single-cell layers at the anterior edge of each parasegment in which *engrailed* gene expression is concentrated after the *eve* and *ftz* stripes have become established. Adapted from Lawrence (1992).

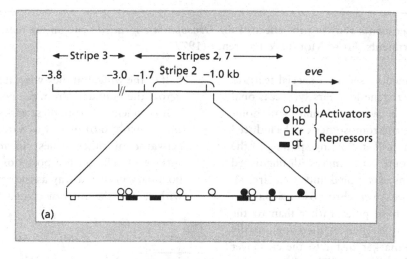

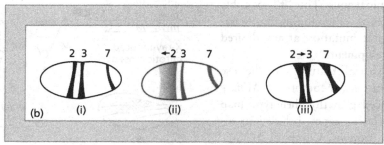

Fig. 7.8 (a) Part of the controlling region of the *Drosophila even-skipped* (*eve*) pair-rule gene. The effects on transcription of separate parts of the transcription-control region upstream of *eve* were tested by fusing them to *lacZ*, and introducing the constructs into flies by P-mediated transformation. The 1.7-kb region upstream of the transcription start-point is sufficient to activate *lacZ* expression in stripes 2 and 7; a deletion of the section between −1.1 and −1.6 kb abolished expression in stripe 2. Inclusion of the region between −3.0 and −3.7 kb extended *eve* expression to stripe 3. The multiple binding sites for *bcd*, *hb*, *Kr* and *gt* protein products are in the positions shown.

(b) (i) β-Galactosidase expression in stripes 2, 3 and 7 induced by a DNA construct linking 5.2 kb of upstream *eve* sequence to *lacZ*. (ii) Effect of disruption of all three of the *gt* protein binding sites shown in (a). Stripe 2 is expanded in an anterior direction. (iii) Effect of point mutation in all five *Kr* protein-binding sites. Stripe 2 is expanded slightly in a posterior direction. Compare with Fig. 7.6. Based on data from Stanojevic *et al.* (1991).

bcd and/or *hb* proteins, which are present at suitable concentrations in the anterior-to-middle part of the embryo, but confined to a narrow stripe by repression exercised by the flanking *gt* and *Kr* products. This hypothesis is confirmed by experiments on *eve-lacZ* constructs in which the putative repressor sites were selectively disrupted. The effect was to allow expansion of stripe 2 *lacZ* expression – in an anterior or posterior direction depending on whether it was the *gt* or *Kr* protein binding sites that had been eliminated (Fig. 7.8). Presumably the *eve* control regions governing the other stripes respond to other combinations of *gap* gene products.

The positioning of the other pair-rule products, including *ftz*, appears to follow from the primary *eve* pattern. The alternating stripes of *eve* and *ftz* expression lead in turn to localized expression of another gene, engrailed (*en*), which was originally recognized by the homeotic phenotype of its homozygous mutant alleles. Homeosis (an 'o' deleted from homoeosis in deference to American usage) means the transformation of one part of the body into another; in the case of *engrailed* mutants the transformation – most clearly seen in the wing – of the posterior halves of segments to the pattern of the anterior halves. The marking of cell lineages by X-ray-induced mitotic crossing

Box 7.2 Experiments showing the effect of the *Drosophila engrailed* gene on the maintenance of tissue compartments. From Morata & Lawrence (1977)

. . . .

These experiments used two special features of *Drosophila* genetics. Firstly, occasional crossing over between chromatids of homologous mitotic chromosomes can be induced by X-irradiation. If, following splitting of the two chromosome centromeres (distinguished in the diagrams as stippled and open circles), the two crossed-over chromosomes pass to opposite anaphase poles rather than to the same pole (a 50% chance) then any initially heterozygous marker distal to the crossover will become homozygous. Thus it is possible to generate clones of cells exibiting the phenotype of a recessive mutation at any desired time during development.

The second special feature was the use of *Minute* mutations. Mutants of *Minute* (dominant reduced growth) phenotype map in numerous different genes, believed to encode ribosomal proteins. Clones of M^+/M^+ cells generated in M/M^+ heterozygotes have a growth advantage and tend to increase their share of developing tissues as cell division proceeds. In the experiments illustrated, two different *Minute* mutations were used, situated in each case between the centromere and one or two recessive markers – *mwh* in the first experiment and *en* and *pwn* in the second. The point of *mwh* and *pwn* is that, when homozygous, they affect the pattern of bristles on wing cells and so provided a visible marker for the M^+/M^+ sectors.

In the first experiment (a) M^+/M^+ *pwn/pwn* clones (shaded) were strictly confined to one or other side of an otherwise invisible straight parasegment boundary. In an *en/en* fly, this boundary was not observed.

In the second experiment (b), the M^+/M^+ clones were homozygous *en/en* in a wing that was otherwise *en$^+$/en*. These clones were restricted in their expansion by the parasegment boundary when they came from the anterior half of the wing, but they were able to expand across the boundary from the posterior side.

It is concluded from these experiments that *en$^+$* is involved (probably as a transcriptional activator of other genes) in marking the surface of cells on the posterior side of the boundary in such a way as to prevent mixing with cells of the anterior compartment.

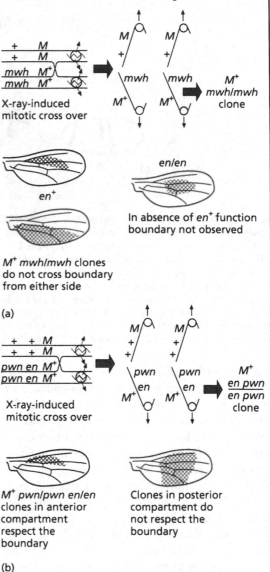

M$^+$ mwh/mwh clones do not cross boundary from either side

In absence of *en$^+$* function boundary not observed

(a)

M$^+$ pwn/pwn en/en clones in anterior compartment respect the boundary

Clones in posterior compartment do not respect the boundary

(b)

over (the principle of which is explained in Box 7.2) suggested that the function of en^+ is to maintain a parasegment boundary which divides a segment into two compartments. Marked clones of wing cells, for example, are usually confined strictly to one side of a line running down the centre of the wing. But in *en/en* homozygous mutants this restriction seems to be abolished. In *en/en*$^+$ heterozygotes, clones that had become homozygous for *en* as a result of induced mitotic crossing over were able to cross from the posterior into the anterior compartment, though not in the reverse direction. More details of these experiments are explained in Box 7.2. It would seem that cells at the anterior boundary of the posterior wing compartment are normally marked by an en^+ product that prevents them from mixing with the cells of the anterior compartment. In accordance with its postulated boundary-observing role, the en^+ protein product is expressed very precisely at the anterior edge of both *eve* and *ftz* stripes (Fig. 7.7).

Much current thinking about *Drosophila* development is based on the concept of compartments. It seems that any cell newly arising by mitosis in a particular compartment acquires a compartmental identity. Each compartment provides a unique environment for the expression of selector genes, to be considered next.

This account of early *Drosophila* development has been highly selective. No mention has been made of the genes controlling the terminal anterior and posterior structures falling outside the parasegments, nor of the genes governing differentiation along the vertical dorsal–ventral axis. We have considered only the development of the obviously segmented ectodermal structures, and given no consideration to the roles of parasegments and compartments in the formation of internal tissues such as mesoderm (especially muscle), which develop after the invagination of the blastoderm to form the gastrula (Lawrence, 1992). The purpose has been merely to show how the first steps of *Drosophila* development, up to the establishment of the parasegments, can be explained in terms of a cascade of transcriptional control, set in train by an initial asymmetry in the egg and made possible by the versatility and subtlety of protein–DNA interactions.

Selector genes in *Drosophila* and other organisms

Drosophila

Selector genes in *Drosophila* were identified through homeotic mutants, in which one compartment of the fly assumed a pattern of development that properly belongs to another. The genes concerned are believed to have the function of selecting (switching on), either directly or indirectly, the whole set of gene activities required for normal development within a particular compartment. In a homeotic mutant the switch is faulty and the wrong set of genes is turned on in the compartment concerned.

The most intensively studied *Drosophila* selector genes are those of the *bithorax* complex, which controls gene expression in the segments of the thorax and abdomen, and the *Antennapedia* complex, which controls the head segments. The genes in both these complexes encode proteins with homeobox sequences, which are characteristic of a certain class of DNA-binding proteins. These proteins are probably all selective transcriptional activators.

The bithorax complex (B-C), on which we will concentrate, includes three genes, *Ultrabithorax* (*Ubx*), *abdominal-A* (*abd-A*) and *Abdominal-B* (*Abd-B*) – the choice of capital versus lower-case initial letters merely reflects the dominance or recessivity of the first mutant alleles to be identified. The three genes extend over about 200 kb of DNA, much of it accounted for by some very long introns as shown in Fig. 7.9. The figure also shows the location of some of the homeotic mutations and their phenotypic effects.

The most immediately informative mutations are those that eliminate gene function altogether. If all three genes are silenced the effect is to convert the third thoracic segment (T3) and eight abdominal segments (A1 to A8) to likenesses of the second thoracic segment (T2). It is as if the pattern of differentiation of T2 is a fall-back plan which will be adopted if the B-C genes convey no instructions to the contrary. The presence of Ubx^+ without $abd\text{-}A^+$ or $abd\text{-}B^+$ allows normal development of T3 and A1, but leaves all the other abdominal segments looking like A1. For correct

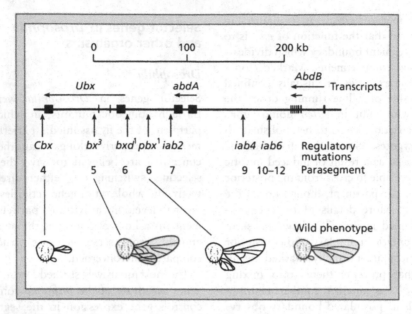

Fig. 7.9 The bithorax gene complex in *Drosophila melanogaster*. Transcription units are shown as arrows and exons as black rectangles or vertical lines. The positions of several homeotic mutations are shown, together with the parasegments (numbered from front to back, see Fig. 7.10) that they affect. The wing and haltere (vestigial 'balancer' wing) phenotypes of some of the mutants are shown in comparison with wild type. The dominant *Contrabithorax* (*Cbx*) shows conversion of wing towards haltere, *bithorax* (*bx*) converts the anterior half of the haltere towards wing, and *postbithorax* (*pbx*) converts the posterior half of the haltere towards wing, with concomitant changes in the fine morphology of the corresponding part of the thorax in each case. Note that the homeotic mutants all fall in regions with transcription-regulation functions outside the exons, and that the sequence of mutational sites along the chromosome mirrors the sequence of the parasegments that they affect (Bender *et al.* 1983).

development of the posterior abdominal segments, the other two B-C genes are required; *abd-A*$^+$ by itself rescues A2 and A3 but leaves A4 to A8 looking like A3, and *abd-B*$^+$ by itself looks after A5 to A8 but does nothing for A2 and A3. These different spheres of influence of the B-C genes agree well with their regions of transcription, as shown in Fig. 7.10. They are all expressed in A5 to A8 of the segmenting gastrula, and *Abd-B*$^+$ transcription is confined to these segments; transcription of *abd-A*$^+$ also occurs in A2 to A4, while that of Ubx$^+$ extends forward as far as T3.

Obviously, the B-C genes must be subject to transcriptional control by segment or compartment-specific activators and/or repressors, and one would expect each such factor to interact with specific DNA sequences. Considerable progress has been made towards the identification of the regulatory DNA sequences using hybrid *Ubx/lacZ*

constructions introduced into embryos with a P-element vector (see Fig. 4.15). There seem to be two regulatory regions for *Ubx*, each 20 kb or more in length, one upstream of the transcription start and one in the long intron. Because of the downstream position of the latter region, and the wide separation of both regions from the transcription start-point, they are appropriately referred to as enhancer rather than promoter sequences. Within these extensive regulatory regions there are shorter sequences apparently concerned with expression in specific parasegments (Fig. 7.9).

The selector genes of the B-C both select, and are themselves selected. For a better understanding of how they work it will be necessary to identify the other genes to which they connect in the sequence of transcriptional controls. On the one hand, we would like to know what gene products

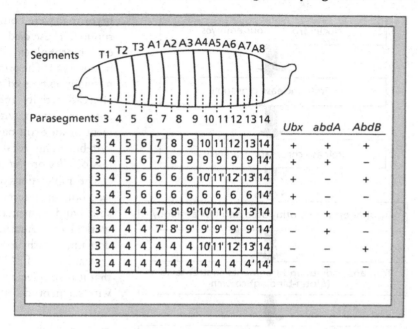

Fig. 7.10 Domains of expression of genes of the bithorax complex, and the parasegmental transformations seen in loss-of-function single, double and triple mutants. The three genes act in a combinatorial fashion; the normal development of each parasegment from 5 backwards requires the appropriate combination of these gene functions. Transformations are indicated by the altered parasegment numbers in the chart. Primed numbers signify that transformation was imperfect, presumably because the parasegments concerned had incomplete sets of gene functions. After Lawrence (1992).

bind to promoter/enhancer sequences of the B-C genes so as to regulate their transcription. On the other hand, we need to find out what other genes are in turn controlled by the B-C products, and what these other genes do. A way of investigating the latter question is sketched in Fig. 7.11.

A remarkable and still mysterious feature of the B-C is that the regulatory elements, both within *Ubx* and in the complex as a whole, occur in a sequence in the chromosome that reflects the posterior-to-anterior order of the larval parasegments that they control (Fig. 7.9). This intriguing correlation is also seen in the *Antennapedia*-complex, some distance away on the same chromosome, and it has parallels in other animals including mouse, as we see below.

Mouse

The discovery of the homeobox domain in *Drosophila* homeotic genes prompted a search for similar sequences in other organisms. The most notable success has been in the mouse. Probing with the *Drosophila Antennapedia* homeobox sequence has revealed four clusters of homeobox-containing genes, called *HoxA, B, C* and *D*, with

up to 13 separately transcribed genes in each cluster (Krumlauf, 1993). In general, the *Hox* genes resemble each other only in their homeobox domains. Even the homeoboxes show much variation, though with recurring motifs on the basis of which they can be classified into families. The same set of families, though with some gaps, are found in each cluster. The similarity between clusters in the ordering of the different families (Fig. 7.12) suggests that the clusters may all have had a common origin. There is also a rather amazing similarity ('parology'), both in homeobox sequences and in gene order, with the homeobox genes of the *Drosophila* bithorax and *Antennapedia* complexes.

Specific nucleic acid probes are available for several of the mouse *Hox* genes. Their transcripts have been identified in various parts of the mouse embryo but especially in the developing vertebral column and brain – the most obviously segmented features. Each gene is expressed in several adjacent prevertebrae, with a well-defined anterior limit. The most striking thing about the pattern of expression is that the order of the genes with respect to their anterior limits of transcription is correlated with their order in the gene cluster (see Fig. 7.12

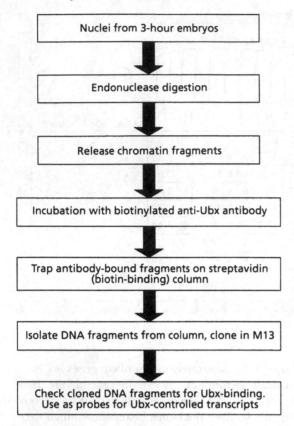

Fig. 7.11 Outline of a strategy for cloning fragments of the control regions of genes subject to control by *Ubx*. Selection is made for DNA fragments bound to *Ubx* protein. After Gould *et al.* (1990).

for more detail). This is very reminiscent of the similar correlation in the *Drosophila* bithorax and *Antennapedia* complexes (BX-C and ANT-C), and its mechanistic basis is similarly mysterious.

Recent studies (Le Mouellic *et al.*, 1992) of the effects of eliminating *Hox* gene function have strengthened the analogy with *Drosophila* homeotic genes. Mouse embryonic stem cells in culture were transformed with a DNA construct carrying *HoxC-8* disrupted by insertion of *lacZ*, and selection was made (for the method, see Fig. 3.13b) for clones in which the disrupted *HoxC-8/lacZ* compound had replaced the normal gene. Such cells were introduced into embryos to make chimaeric mice, one of which transmitted the disrupted gene through the germ line. A further generation of inbreeding led to the production of em-

bryos that were homozygous for the *HoxC-8* disruption. These died early in development, but the early morphological effects of the disruption could be observed. The prevertebrae in which *HoxC-8* is normally expressed now showed *lacZ* instead of *HoxC-8* activity and were converted to a more anterior pattern, with, for example, the development of an extra pair of ribs on the first lumbar vertebra. This recalls the anteriorizing effect of loss of *Ubx* or *abd-A* function in *Drosophila*.

The implications of these almost uncanny parallels, both in insects and mice, between chromosomal and segmental organization have still to be worked out. Another mystery is the homeobox domain, which appears to be just one among several kinds of DNA-binding structure. Why is it that it is so often associated with genes concerned with segment identity?

Flowering plants

In two flowering plant species, *Arabidopsis thaliana*, and *Antirrhinum majus* (snapdragon), intensive mutant hunts have produced very useful collections of homeotic mutants, opening up the prospect of a molecular genetics of development of flowering plants parallel to that of *Drosophila*.

The typical flower consists of four whorls of organs – sepals, petals, stamens and carpels – in threes, fours, fives or higher numbers according to the plant family. Our two examples both have flower structures in fives, with some distortion in the asymmetric snapdragon. The different mutants can be grouped into three different classes, A, B and C. Classes A and C each include only one gene in each species so far, though more may well be discovered; three known genes in snapdragon and two in *Arabidopsis* fall into class B. Triple mutants homozygous for null alleles of genes in all three classes (a^- b^- c^-) produce leaf-like structures in place of the normal organs in all four whorls. The A, B and C functions act in combinatorial fashion to produce the four kinds of organ in the proper positions: C alone produces carpels, A alone sepals, C together with B petals and A together with B stamens. The phenotypes are sketched in Fig. 7.13, and the interpretation of the gene effects is explained in Fig. 7.14. The clear prediction is

Fig. 7.12 The arrangement of the *Hox* genes of mouse in four unlinked clusters *A–D*. The genes within each cluster are of the order of 1–2 kb apart and are drawn evenly spaced in their correct linkage order. Genes in the same vertical row have very similar homeobox sequences and for this reason are placed in the same gene family (families numbered 1–13). On the same basis, several families can be related to *Drosophila* homeotic genes (foot of diagram). The upper diagram shows the zones of expression – sharply delimited at the anterior end – of members of the *HoxB* cluster in the mouse embryonic hind brain and vertebral column. The other clusters show a similar correlation between map position and zone of expression. After McGinnis & Krumlauf (1992) and Krumlauf (1993).

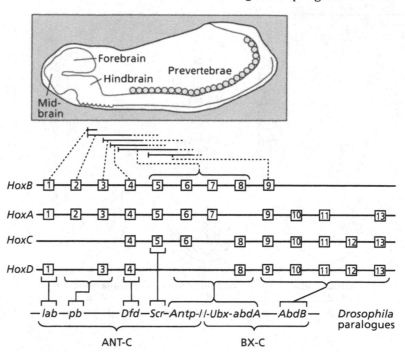

that each gene is selectively transcribed in two adjacent whorls in the flower primordium. This has been confirmed so far as probes for the gene transcripts are available. A study of the *Antirrhinum* gene class C gene *plena*, for example, showed that it was transcribed in the wild type only in the two inner whorls (carpels and stamens). A null *plena* mutant has these organs converted to petals and sepals, while a deregulated *plena* allele which was transcribed in all four whorls converted sepals and petals to carpels and stamens (Bradley *et al.*, 1993).

The A, B, C classes of flowering plant organ identity genes seem closely analogous to the *Drosophila* selector genes, acting singly or in combination to select one or other of a number of alternative developmental subprogrammes (Fig. 7.14). However, the representatives sequenced so far do not encode homeodomains but instead have another kind of DNA-binding domain called, for rather esoteric reasons not worth explaining, the MADS box.

At the time that the A, B and C genes have their effects, the stage has already been set for their action; each developing floral meristem is already committed to producing four fivefold whorls of structures of some kind or other – leaves in the absence of further instructions. In snapdragon there is, so far, one gene known to be necessary for setting this pattern. The *floricaula* mutant, which is epistatic to the organ identity genes, produces no flower-like whorls at all. It does develop arrays of meristems (clusters of actively dividing cells) arranged as if they were going to yield flowers, but instead they all grow into ordinary leafy shoots. This may be just the first of a number of discoveries of genes involved in the earlier steps of the flowering programme.

The picture that emerges, though much less detailed than for *Drosophila*, is quite similar in essence. Morphological structure emerges in stages. At each stage, interacting sets of genes build upon and are constrained by patterns set by earlier-acting sets.

Timing of gene expression – a function for introns?

An aspect of gene expression that has only recently been given attention is the lapse of time between

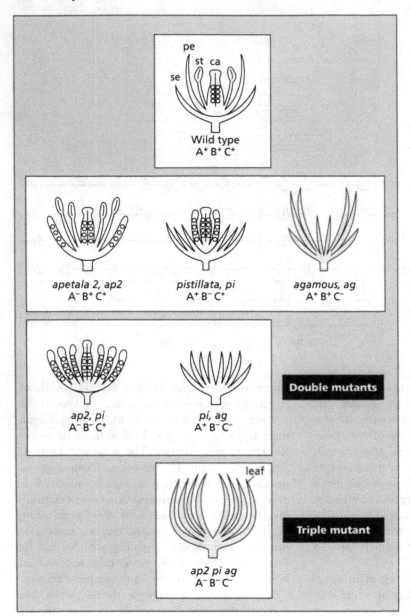

Fig. 7.13 (a) Effects of some homeotic gene mutations in the flowering plant *Arabidopsis thaliana*. Sepals (se), petals (pe), stamens (st) and carpels (ca) are symbolized as indicated in the wild-type diagram at the top. Only two of the five structures present in each whorl are visible in the diagrammatic vertical section. The mutations are all recessive and the phenotypes shown are of homozygotes. In the homozygous triple mutant all four whorls of organs are transformed into leaf-like structures. The *ap2, pi* and *ag* mutants are representative of mutant groups A, B and C (see Fig. 7.14). After Coen & Meyerowitz (1991).

the initiation of transcription and the appearance of the protein product. RNA synthesis has been estimated to proceed in *Drosophila* at a rate of 1.1 kb per minute. This gives introns a significance that has been generally overlooked. A transcription unit padded out with 50 kb of introns, as is not at all unusual in organisms with complex modes of development, will take more than half an hour to express itself in terms of protein product from the time of initiation of transcription. A positive transcriptional signal acting on a long transcription unit is a sort of time-switch, set to produce an effect at some fairly precisely defined future time.

The importance of delay in gene expression is shown by some recent work on a *Drosophila* gene *knirps-like* (*knrl*), which is closely linked to *knirps* (*kni*, see pp. 196–197) and encodes a rather

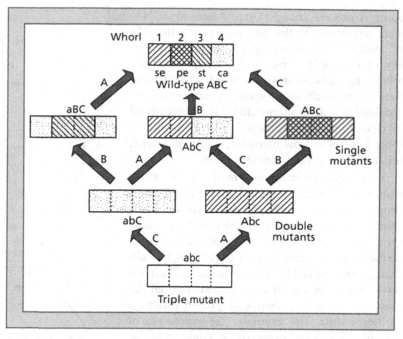

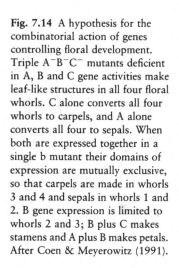

Fig. 7.14 A hypothesis for the combinatorial action of genes controlling floral development. Triple $A^-B^-C^-$ mutants deficient in A, B and C gene activities make leaf-like structures in all four floral whorls. C alone converts all four whorls to carpels, and A alone converts all four to sepals. When both are expressed together in a single b mutant their domains of expression are mutually exclusive, so that carpels are made in whorls 3 and 4 and sepals in whorls 1 and 2. B gene expression is limited to whorls 2 and 3; B plus C makes stamens and A plus B makes petals. After Coen & Meyerowitz (1991).

similar protein with virtually identical DNA-binding specificity. The most notable difference between the two genes is their intron content – *kni* has about 1 kb of intron sequence compared to 19 kb in *knrl*. The *kni* gene product is needed for the laying down of pattern at a time when the nuclei are dividing every few minutes to form the blastoderm. Various engineered derivatives of *knrl* have been tested for their ability to rescue homozygous mutant *kni* embryos. A compact derivative made from cDNA and thus devoid of introns was found to support relatively complete embryonic development, with restoration of most of the structures lost through lack of *kni* function, even though it did not restore viability. The full-length *knrl* gene, however, was quite ineffective as substitute for *knirps* (Rothe *et al.*, 1992). The likely explanation is that, to be effective, transcription and translation have to be completed between one mitosis and the next, and transcription of wild-type *knrl* just takes too long in mitotically hyperactive blastoderm. Presumably *knrl* performs some other function for which a longer lead time is appropriate.

In principle, combinations of genes with different lead times for their expression could provide for rather precise temporal patterns of gene activity. One can imagine that genes with different time settings, activating or repressing each other's transcription, could produce pulses or cycles of activity on a time scale of minutes or hours. Changes in intron length could be a way of fine-tuning the system. Examples of the use of this kind of mechanism are awaited.

Stabilizing gene activity – epigenetic effects

Transcriptional activity/inactivity – mechanisms of maintenance

Different kinds of cells in a multicellular organism generally show very different patterns of gene transcriptional activity, each expressing its own particular subset of genes. These patterns are often rather stable. Even when taken out of the organism and cultured in an artificial medium, different cell types often retain at least some of their differentiating features in culture. It is evident that patterns of gene transcription can show significant heritability through cycles of mitosis. In some instances cell-heritable differences are dependent upon structural changes in the genomic DNA, most famously in the antibody-producing cells of vertebrate immune systems but also in a variety of reversible switches of cell type in single-celled organisms – protozoa, yeasts and bacteria. But, in multicellular organisms, most stable or quasi-stable cell differentiation is not underpinned by changes in DNA sequence and must be due to some kind of self-maintenance, called *epigenetic*, superimposed on the constant DNA.

In principle, there are two main ways in which epigenetic differences could be maintained. Most simply, it could be through positive feedback of a direct or indirect product of the activity of a gene on its own transcription. One case in point, described earlier in this chapter (p. 188), is the self-maintenance of *GAL1* activity in yeast. There are now numerous examples of transcriptional activation of genes by their own protein products. In *Drosophila*, for example, the expression of *engrailed* has been shown to be stabilized in this way.

Maintenance of activity by positive feedback need not involve any special structure of the chromatin. There are nevertheless a great many demonstrations of correlation between the state of activity of genes and the degree of chromatin compaction as measured by protection of the DNA from digestion by DNase. The DNA of actively transcribed chromatin is characteristically sensitive to DNase digestion, whereas transcriptionally inactive DNA tends to be more effectively protected by bound histones and other proteins. The greater compaction of the transcriptionally silent heterochromatic regions of chromosomes can be visible under the microscope. Transcriptional activity is also correlated with DNA under-methylation and, as we see below (see Fig. 7.17), with histone acetylation.

How might states of transcriptional activity or inactivity of chromatin be transmitted through cycles of replication? The only precisely defined model depends on maintenance of methylation. There is evidence that at least one of the enzymes that methylate DNA acts preferentially on half-methylated sites – sequences (5'CpG3' in animals and 5'CpXpG3' in plants) that have the cytosine methylated in one strand but not in the other. Such an enzyme will presumably methylate the half-methylated DNA resulting from replication

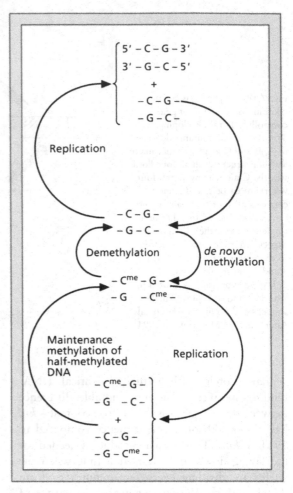

Fig. 7.15 Scheme for the maintenance of methylation of DNA through cycles of replication. Maintenance methylase specifically methylates half-methylated DNA, specifically (in animals) in C residues 5′ to G. Each replication of fully methylated DNA generates half-methylated DNA for the maintenance enzyme to act upon. Switches from the methylated to the non-methylated state, and *vice versa*, may depend on special demethylating and *de novo* methylating enzymes. After Holliday & Pugh (1975).

of fully methylated DNA, but will fail to methylate wholly unmethylated DNA. This is a mechanism for maintenance but not for gain or loss of methylation, for which *de novo* methylating or demethylating enzymes must be invoked (Fig. 7.15).

Whether this is an adequate model for epigenetic effects in general is very doubtful, if only because *Drosophila* and fungi normally have little or no

DNA methylation, whether genes are transcribed or not. In fact, the correlation between methylation (where it exists) and transcriptional inactivity of DNA can be interpreted in more than one way. It may be that methylation is a symptom rather than a cause of transcriptional inactivity, rather as an unused machine tends to gather rust. Or perhaps methylation is a contributory, but not the sole or main cause. A plausible though imprecise hypothesis is that the difference between transcriptional activity and inactivity depends on the maintenance through cycles of replication of particular DNA–protein complexes. This idea is discussed below (p. 215).

X-chromosome inactivation in mammals

The Y-chromosome of male mammals carries information essential for maleness, but few if any genes with functions common to males and females. The X-chromosome, on the other hand, carries many genes with functions that are equally important for both sexes. How, then, does the male make do with only one dose of all the X-linked genes when the female has two? The answer is that the dosage is effectively single in both sexes, since one of the two Xs in the female mammal is inactivated epigenetically early in embryonic development. The inactivation occurs in every female cell and effects either X with equal probability, so that the developed female is actually a mosaic, with one X active in about half the cells and the other X active in the other half.

The mosaicism can be most easily seen in females heterozyous with respect to a pair of X-linked alleles affecting coat colour. To take the best-known example, the black/ginger difference in cats is determined by the X-linked alleles *B/b*. Tom cats may be either *B* or *b*, black or ginger. Females are usually either *B/B* (black) or *B/b* (tortoiseshell). The tortoiseshell pattern is a mosaic of clones of cells expressing *B* or *b*, depending on whether the *b*- or the *B*-carrying X is inactivated (the white patches frequently seen are an effect of a dominant allele of another gene).

In female mammals, including human females, the inactive X in every cell nucleus is visible as a compact darkly-staining blob or *Barr body*, named after the Canadian cytologist who first described it. Its permanent compaction and transcriptional inactivity are maintained indefinitely through cycles of replication in somatic cells, but not in the germ line, which gives rise only to eggs with active X-chromosomes. In those rare individuals, whether mice or humans, that have three X-chromosomes, there are two Barr bodies – two Xs inactivated and one remaining active. The mechanism that counts the X-chromosomes and inactivates all except one on a random basis is as yet unknown. Interestingly, in marsupials, inactivation is not at random as it is in ordinary mammals, but affects always the paternal X. This implies that the X-chromosomes must retain some kind of epigenetic 'memory' of parental origin – an idea that recurs below (p. 231) in our discussion of the 'imprinting' phenomenon in mouse.

Studies of autosome-X-chromosome rearrangements in mouse have shown that at least some parts of the X-chromosome can transmit their state of inactivation to any autosomal fragment to which they become joined. When a piece of an autosome is inserted into the X (as in Cattenach's translocation, see Fig. 7.16) inactivation can spread into it from both ends, with genes in the middle of the autosomal fragment most likely to remain active. The intriguing thing about this phenomenon is that different cell clones show different extents of inactivation. This patchy, clonal pattern is also seen in *Drosophila* position effect variegation (see below).

The compact molecular structure of the inactive X is probably similar to that of heterochromatin in general and stabilized by a special set of proteins, mostly as yet uncharacterized. Inactivation is certainly accompanied by increased DNA methylation (Riggs & Pfeifer, 1992) and, as recently shown, by a lack of acetylation of the core histone H4, a property shared with heterochromatic segments of the autosomes (Fig. 7.17a).

Dosage compensation in *Drosophila*

Flies have the same problem as mammals in making the effective X-chromosome dosage the same in females as in males, but they solve it in quite a different way. Instead of inactivating one X in the female, they contrive to make the single X in the male work twice as hard as each female X. This

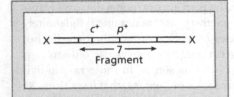

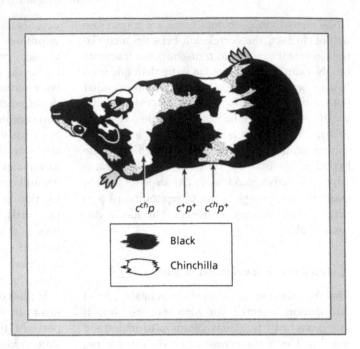

Fig. 7.16 Coat colour variegation in a female mouse with an insertion into one X-chromosome of a fragment of chromosome 7. The mouse has two structurally normal 7th chromosomes homozygous for the recessive coat colour mutants chinchilla (c^{ch}) and pink-eye (p), the latter interacting with c^{ch} to give white hair. The insertion carries the wild-type alleles c^+ and p^+, with c^+ near one end and p^+ in the middle of the fragment. Inactivation of the X chromosome with the insertion usually results in inactivation of the inserted c^+, but the zone of inactivation spreads as far as p^+ only in parts of the c^{ch} cell clones. After Cattenach (1974).

can easily be demonstrated by comparing the effect of a weakly active X-linked mutant allele when it is present in the single male chromosome with its effect when in heterozygous combination with a null allele in the female. A good example is the *apricot* allele of the *white* gene (relevant alleles: w white, w^a apricot, w^+ red eye). Males of constitution wa/Y have about twice as much pigment in their eyes as w^a/w females.

The *Drosophila* male X-chromosome is, at least in the polytene nuclei of the salivary glands, strikingly different both from female Xs and from autosomes of both sexes in its pattern of acetylation of histone H4. Fluorescent antibodies specific for H4 molecules acetylated at lysine residues 5, 8, 12 or 16 were used to show that, while acetylation at the first three positions occurred generally on all chromosomes, lysine-16 was appreciably acetylated only in the male X (Fig. 7.17b).

Thus histone H4 acetylation, at least at lysine-16, correlates with transcriptional hyperactivity in *Drosophila*, just as under-acetylation correlates with inactivity in the female mammalian X. In both cases, the state of activity is maintained through cycles of replication – either mitosis or polytenization.

Drosophila position–effect variegation and the maintenance of states of gene activity during development

It has been known for over 50 years that *Drosophila* genes can be reversibly inactivated when placed close to heterochromatin through segmental chromosome interchanges or inversions. This kind of inactivation is usually sporadic, affecting some cells but not all, and appears to be heritable through cycles of cell division to give patches, apparently clones of cells, showing either wild or mutant phenotype.

Gene inactivation spreads progressively from heterochromatin into the transposed euchromatic segment. The greater the distance of the gene from the euchromatin–heterochromatin junction, the lower the probability that it will be inactivated. If, for example a euchromatic chromosome segment carrying several genes is transposed into a block of heterochromatin, the genes at the ends of the inserted segment may show variegated expression while the ones in the middle, relatively far from heterochromatin, are uniformly expressed (Fig. 7.18; compare the effect of insertion into the mouse X-chromosome, Fig. 7.16).

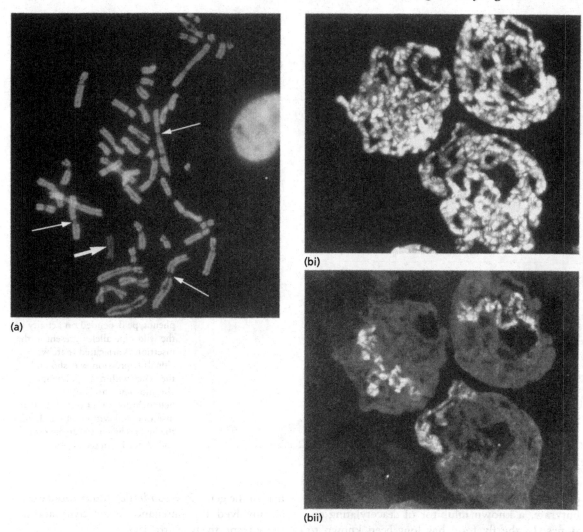

Fig. 7.17 (a) A metaphase chromosomes spread from a human female stained with fluorescent antibody against acetylated histone H4. The inactive X (large arrow) has virtually no acetylated H4 and is visible only through the general orange background stain (shown by large arrow). The antibody gives a banding pattern in the other chromosomes similar to R-banding (compare G-banding, Fig. 5.7). Note the lack of H4 acetylation in centromeric heterochromatin (smaller arrows). From the colour photograph of Jeppesen & Turner (1993). Courtesy of Dr P. Jeppesen.

(b) A male *Drosophila melanogaster* polytene nucleus stained (i) with a general fluorescent stain for chromatin (Hoechst) and (ii) with fluorescent antibody specific for histone H4 acetylated at lysine-16, which shows up only the X-chromosome. H4-acetylation at lysine-16 is believed to be part of the mechanism for hyperactivation (dosage compensation) of the single male X. From Turner *et al.* (1992). Courtesy of Professor B.M. Turner.

Various lines of evidence suggest that position–effect inactivation depends on the binding of certain chromosomal proteins at the affected loci (Reuter & Spierer, 1992). Several different genes can mutate to suppress the inactivating effect, and at least one of these, *Su(var)205*, has been shown to encode a protein that is normally associated with heterochromatin. A similar suppression of position–effect variegation can be brought about by extra copies of the Y-chromosome, and this can be plausibly interpreted as due to the sequestration by the extra Y-heterochromatin of heterochromatin-associated proteins that would otherwise be available for inactivation of position-affected genes.

Here again there is evidence for the importance

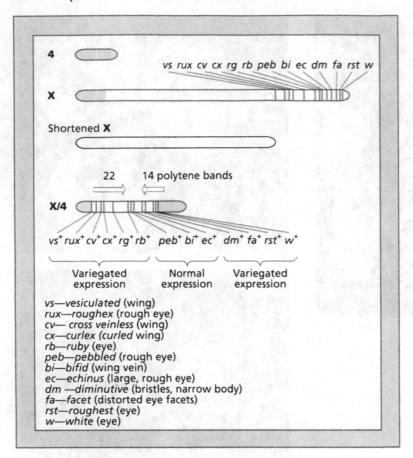

Fig. 7.18 An example of position-effect variegation in *Drosophila melanogaster*. The position-effects were shown by females heterozygous with respect to a chromosome rearrangement in which a distal segment of the X was inserted into the heterochromatin (shaded in the figure) of chromosome 4. The normal X carried recessive mutant alleles of the genes indicated, so the phenotype depended on activity of the wild-type alleles present in the insertion. Variegated (patchy, clonal) expression was shown by the genes within 22 polytene chromosome bands of the heterochromatin at one end of the insertion and within 14 bands of the heterochromatin at the other end. After Demerec (1940).

of acetylation, presumably of histones. Adding butyrate, a known inhibitor of deacetylating enzymes, to the fly food has long been known to counteract position–effect variegation. The likelihood is that a number of different proteins – both non-acetylated histones and non-histones – serve to stabilize a clonally heritable inactive state of chromatin.

The epigenetic stabilization of gene activity or inactivity appears to be important in the normal unfolding of the *Drosophila* developmental programme as well as in position-effect variegation. The selector genes of the bithorax (BX) and Antennapedia (ANT) complexes remain activated in some cell lineages and repressed in others long after the expression of the segmentation genes, that set up these states of activity, has faded away. Evidence is accumulating that two different sets of chromosomal proteins, exemplified by the prod-

ucts of the genes *Polycomb* (*Pc*) and *trithorax* (*tx*), are involved in maintenance of inactivity and activity respectively (Paro, 1990).

The original mutant *Pc* allele is a dominant that causes male-specific pads of hair – *sex-combs* – to develop on all legs instead of just on the front pair. When homozygous the allele brings about a much more drastic and lethal shift to more posterior patterns of segmental development. The genes of the BX-complex are expressed in anterior segments where they would normally be repressed, and all segments tend to the pattern of abdominal-8.

Polycomb encodes a chromatin protein that binds at many different loci throughout the *Drosophila* genome. It is the best studied of a group of genes whose protein products form large aggregates in transcriptionally inactive chromatin (Messmer *et al.*, 1992). Near the N-terminus of the *Pc* polypeptide chain there is a 48-residue sequence that recurs, with surprisingly little variation, in nuclear proteins of diverse organisms. This feature, called the *chromo-domain*, is shared by the protein product of the *Su(var)205* gene, which, as mentioned above, is involved in the clonally transmitted position-effect inactivation that spreads from heterochromatin.

In contrast to the *Polycomb* group of proteins, the *trx* product is believed to be one of several chromosomal proteins needed for maintenance of transcriptional activity. This protein has a zinc-finger domain strongly suggestive of DNA binding. Its total loss by mutation is lethal; alleles with reduced but not zero function give phenotypes consistent with underexpression of the BX and ANT complexes.

Imprinting of mammalian chromosomes

Until recently, it has been axiomatic that, apart from sex chromosomes mitochondrial traits, the genetic contributions of male and female parents in animal crosses were equivalent. Modern techniques of nuclear transplantation, however, have shown that, at least in the mouse, this is not true.

Immediately after fertilization, the mammalian egg contains two haploid pronuclei, one originating with the egg and one contributed by the sperm cell. Newly fertilized eggs can be recovered from a female mouse, and by microsurgery it is possible to remove one of the pronuclei and replace it by another, before reimplanting in the oviduct of another female. When a pronucleus is replaced by another of the same gender (i.e., egg nucleus by egg nucleus, or sperm nucleus by sperm nucleus), embryonic development often proceeds normally, leading to birth of a normal mouse. But when a pronucleus is replaced by one of opposite gender, so that the manipulated egg finishes up with two egg or two sperm nuclei, development is aborted after several cell divisions, though it gets further when its parentage is all-female than when it is all-male.

Haploid nuclei of eggs and sperm are therefore not functionally equivalent in the mouse. Furthermore, their non-equivalence has been shown to persist through several mitotic divisions. If a fertilized egg with one of its pronuclei removed is

cultured under suitable conditions, without receiving a replacement nucleus, it can develop for a limited time as a haploid embryo, called *gynogenetic* or *androgenetic* depending on whether its haploid genome is of female or male origin. A nucleus from such a multicellular haploid can then be used to replace a pronucleus removed from a fertilized egg. The results show that nuclei from an androgenetic embryo up to the four-cell stage (which is as far as most androgenetic embryos develop), or the nuclei from gynogenetic embryos up to the eight-cell stage can, when introduced into eggs with one surviving pronucleus, compensate for the loss of a pronucleus of their own gender only. This shows that the non-equivalence of egg and sperm genomes is inherited through several cycles of haploid mitosis; an effect known as *imprinting*.

The gender-dependent imprinting phenomenon has been linked to differential methylation of stretches of chromosomal DNA. Some genes are characteristically highly methylated in the early embryo when transmitted through the egg but not when transmitted through the sperm; for other genes the situation is the other way round. The first demonstrations of such differences were made using transgenes. For example, in one series of experiments, the bacterial gene coding for chloramphenicol transacetylase (*CAT*), was inserted into the mouse genome (see p. 121) at random locations following injection into an egg pronucleus. Most of these *CAT* insertions were equally methylated in progeny mice regardless of which parent they had come from, but at one particular chromosome locus the *CAT* gene was much less methylated when inherited paternally than maternally. This effect was clearly a function of the position of insertion of the gene, rather than of the DNA sequence of the gene itself. An under-methylated *CAT* copy inherited by a female from the father became fully methylated when transmitted from that female to the next generation; conversely, a fully methylated maternally derived copy became undermethylated again when transmitted by a male. Thus imprinting is reversible; it is reset for each new generation of germ cells.

More recently, the same kind of gender effects have been shown to apply to natural mouse genes. Linked genes can be imprinted in opposite ways,

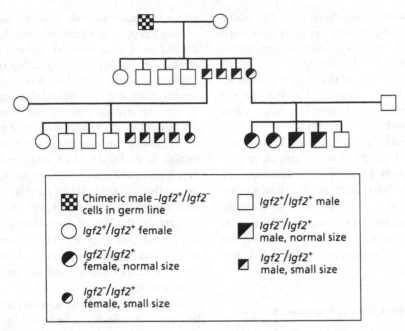

Fig. 7.19 Part of a mouse pedigree, from De Chiara *et al.* (1991), showing the effects of maternal imprinting of *Igf2*, encoding insulin-like growth factor II. Embryonic stem cells were selected for disruption of *Igf2* (for the method see Fig. 3.13b), and mutant cells with one gene copy inactivated (heterozygous *Igf2⁺/Igf2⁻*) were introduced into embryos to make chimeric mice. A male chimera with a predominance of heterozygous mutant cells in its germ line was mated to a normal female. Progeny inheriting *Igf2⁺* from the male parent were normal in size, but those inheriting *Igf2⁻* were small because their maternal *Igf2⁺* allele had been inactivated by imprinting. The progeny from a small male mated to a normal female were normal or small depending on whether they inherited *Igf2⁺* or *Igf2⁻* from their heterozygous *Igf2⁺/Igf2⁻* father — again, *Igf2⁺* from the mother was inactive. Note that the occurrence of normal-size progeny from this cross demonstrates that imprinting is reversible; the grand-maternal *Igf2⁺* allele, inactive during paternal development, is restored to activity in the paternal sperm. Progeny from mating a small *Igf2⁺/Igf2⁻* female to a normal male were all normal in size because they all inherited non-imprinted *Ifg2⁺* from their father.

and so it is clear that the unit that is imprinted must be much smaller than a whole chromosome. Thus the mouse gene *Igf2*, encoding insulin-like growth factor, is transcribed in the embryo only when inherited from the male parent (genetic evidence is shown in Fig. 7.19), whereas in the case of a closely linked gene called *H19* it is the maternally-derived allele that is active. The function of *H19* is not known, but it is strongly transcribed; a difference in transcript length between different *H19* alleles was used to show that only the maternal allele was transcribed in the embryo.

The correlation between imprinting and methylation is not complete. The paternally contributed *H19* allele has been found to be hypermethylated in the early embryo but not in the sperm, so the nature of the imprint in the germ cell is not clear. In the case of *Igf2*, no difference in methylation

has been found between maternally and paternally contributed alleles.

The nature and significance of chromatin structure

Although it is early to draw general conclusions, it is tempting to speculate that all the examples of self-perpetuating states of chromatin activity/inactivity reviewed in this section, as well as several others that could have been mentioned, have a common basis. The way to a better understanding is likely to be through the analysis of mutants such as those of the *Drosophila Polycomb* and *trithorax* groups, as well as comparable mutants that are beginning to be analysed in yeast. We can look forward to the identification of the proteins encoded by the genes concerned, and the study of their roles in chromatin structure and replication.

The demonstration that different chromosomal proteins are associated with different states of chromatin activity does not, in itself, tell us how these states are transmitted through cycles of chromosome replication. There is no protein-based model for epigenetic transmission comparable in elegance to the DNA-methylation hypothesis (see Fig. 7.15), which can hardly apply to *Drosophila*. Nevertheless, the form that a protein model might take can at least be guessed at.

There appear to be two kinds of transmission of inactive states of chromatin: 'horizontally' along the chromosome, as seen in heterochromatin-based position effects, and 'vertically' through cycles of replication, as in X-chromosome inactivation and imprinting. There may be a connection between the two.

Observations of position–effect in *Drosophila* led long ago to the idea of horizontal propagation of heterochromatin. If, as seems increasingly likely, the inactive state of chromatin is stabilized by cooperative binding of an array of different proteins to form an open-ended repetitive complex, it is not difficult to imagine that complex being extended indefinitely so long as the protein building blocks continue to be available. The idea of horizontal transmission of chromatin structure, vague though it still is in molecular terms, may also be relevant to the puzzling correlations in mouse and *Drosophila* between gene order and the anterior–posterior order of the body segments in which the genes are expressed (see Fig. 7.12). One speculation is that a boundary between active and inactive chromatin moves for different distances depending upon some anterior–posterior gradient.

To explain vertical (epigenetic) transmission on the same lines it is only necessary to suppose that the protein complex responsible for chromatin inactivity is divided between daughter chromatids in such a way as to provide each with the minimum units needed to 'seed' further complex formation. Of course, if genes are ever going to be reactivated there must be some way of breaking out of this cycle. Perhaps the modification of one component (possibly histone acetylation) can cause the whole complex to fall apart, to be replaced by a provisionally self-sustaining *active* chromatin complex. Now that some of the proteins involved are becoming known it may be more possible to test these still vague ideas. The general significance of epigenetic mechanisms is that they allow the separation of the maintenance of states of gene activity from the conditions for their initiation. Switching a gene or a block of genes on or off at a particular time and place in development is likely to require more or less transient signals from an array of controlling genes. But once a pattern of transcription, and the chromatin structure associated with it, has been established it may enjoy a considerable degree of autonomy. This is likely to be the main basis of the stable or quasi-stable cellular differentiation that is characteristic of eukaryotic development.

Towards the complete description of the organism

The development of a whole organism from a single cell is a highly deterministic process. Within a species, a detailed developmental programme is played out with great predictability. There is no doubt that the information for the programme is encoded in the genome. It is sometimes implied that, when every gene and open reading frame has been identified, we will have a complete handbook and guide to how the organism develops and functions. Clearly this will not be the case. There will still remain the problem of tracing the connections between the genes and the phenotype, and this is a task of incomparably greater magnitude than the sequencing of all the DNA.

One major problem concerns the many genes that (to generalize from yeast) do not appear to connect to the phenotype at all. As was stated on p. 157, a majority of open reading frames in the DNA of *Saccharomyces* chromosome 3 could be disrupted without evident effect on the functioning of the whole organism; there is no reason to think that this situation will be peculiar to yeast. It is difficult to believe that all those open reading frames could have been conserved if they had no function at all. Perhaps they are needed under special environmental conditions that have not yet been reproduced in the laboratory. Possibly, as some have suggested, a high degree of redundancy in gene function has a stabilizing effect, but there is little clear idea as to how this might work.

For many genes, of course, function is obvious. The organism, after all, is built on the basis of proteins – both structural proteins and proteins acting as enzymes – the amino acid sequences of which can in principle all be directly related to the genes that encode them. But for normal metabolism, and even more for the assembly of cells and higher-order morphological structure, the expression of each gene has to be controlled in quantity, space and time. This control is, as we have seen, exercised by other genes, which are themselves subject to control. The genes do not determine the phenotype through a series of separate connections like telephone lines, but rather through a network or mesh of interactions. The branches in the network are both converging and diverging, in the sense that one gene or gene product can be subject to multiple simultaneous controls, while a single gene may participate in the control of several others.

We have seen how, in the paradigm case of the *Drosophila* embryo, segmental structures are built up step-by-step, with each step setting the stage for the next. At each step the pattern of interactions changes. It is, so to speak, a highly complicated game that generates its own rule changes as it goes along. Moreover, a gene that plays a particular role at one stage may do something quite different at another.

So what are the chances of developmental geneticists and biochemists being able to unpick, piece by piece, the entire network connecting genotype and phenotype? Connections can be traced starting from the gene sequences and moving 'outward' to their consequences, or starting from the macromolecular structures of the developed organism and moving 'inward' toward their genetic causation. Current methodology allows the demonstration of several kinds of link. Knowing the identity of a promoter or enhancer, one can find out what proteins bind to it, and what genes they are encoded by. Knowing a DNA-binding protein, one can look for its DNA target(s). If a control gene encodes what looks like a protein kinase, the protein(s) on which it acts and their encoding genes can in principle be identified – and so on. At each step one can see what

particular aspects of the phenotype a particular bit of the network ultimately connects to by mutating each of the relevant genes and observing the effects.

The basic problem of biology is its immense complexity. However complicated the map that is made of the controlling network, there may always be unidentified links and loose ends. Progress is made *towards* the complete description of the organism, but there is no assurance that the goal will be reached. Does this vitiate the whole enterprise? There is a point of view that a living organism is such a highly integrated system that no part of it can be fully understood without a full description of the whole. In a rather pedantic sense, this is no doubt true, but the examples cited in this chapter, not to mention the successes of biotechnology and molecular medicine, show that it is possible to achieve a useful and reasonably satisfying degree of understanding of many aspects of living systems. There are, however, always likely to be undiscovered connections, and some of these may be important in surprising ways.

Further reading

Genetic control of development
Wilkins, A.S. (1986) *Genetical Analysis of Animal Development*. John Wiley, New York. (A general background to this chapter.)

Programmed DNA arrangements
Trends in Genetics (1992) 8, 403–61. Special Issue on Programmed DNA Rearrangements. (This reference deals with special mechanisms of cell differentiation, in the vertebrate immune system and various unicellular organisms, that are not dealt with in this book.)

References

Baker, B.S. (1989) Sex in flies: the splice of life. *Nature*, 259, 471–9.

Bender, W., Akam, M., Karch, F. *et al.* (1983) Molecular genetics of the bithorax complex in *Drosophila melanogaster*. *Science*, 221, 23–9.

Bhat, P.J. & Hopper, J.E. (1992) Overproduction of the GAL1 or GAL3 protein causes galactose-independent activation of the GAL4 protein; evidence for a new model for induction for the yeast *GAL/MEL* regulon. *Mol Cell Biol*, 12, 2701–7.

Bradley, D., Carpenter, R., Sommer, H., Hartley, N. & Coen, E. (1993) Complementary floral phenotypes result from opposite orientations of a transposon at the *plena* locus of *Antirrhinum majus*. *Cell*, **72**, 85–95.

Cattenach, B. (1974) Position-effect variegation in the mouse. *Genet Res Camb*, **23**, 291–306.

Cigan, A.M., Bushman, J.L., Boal, T.R. & Hinnebusch, A.G. (1993) A protein complex of translational regulators of *GCN4* mRNA is the guanosine nucleotide-exchange factor for translation initiation factor 2 in yeast. *Proc Natl Acad Sci USA*, **90**, 5350–4.

Coen, E.S. & Meyerowitz, E.M. (1991) The war of the whorls: genetic interactions controlling flower development. *Nature*, **353**, 31–7.

De Chiara, T.M., Robertson, E.J. & Epstratiadis, T.M. (1991) A growth deficiency phenotype in heterozygous mice carrying an insulin-like growth factor II gene disrupted by targetting. *Cell*, **64**, 849–59.

Demerec, M. (1940) Genetic behavior of euchromatic segments inserted into heterochromatin. *Genetics*, **25**, 618–27.

Dever, T.E., Feng, L., Wek, R.C. *et al.* (1992) Phosphorylation of initiation factor GCN2 mediates translational control of *GCN4* in yeast. *Cell*, **68**, 585–96.

Forsberg, S.L. & Nurse, P. (1991) Cell cycle regulation in the yeasts *Saccharomyces cerevisiae* and *Schizosaccharomyces pombe*. *Annu Rev Cell Biol*, **7**, 227–56.

Gould, A.P., Brookman, J.J., Strutt, D.I. & White, R.A.H. (1990) Targets of homeotic gene control in *Drosophila*. *Nature*, **348**, 308–12.

Hinnebusch, A.G. (1988) Mechanisms of gene regulation in the general control of amino acid biosynthesis in *Saccharomyces cerevisiae*. *Microbiol Rev*, **52**, 248–73.

Hinnesbuch, A.G. (1993) Gene-specific translational control by phosphorylation of eukaryotic initiation factor 2 (Review). *Molec Microbiol*, **10**, 215–25.

Holliday, R. & Pugh, J.E. (1975) DNA modification mechanisms and gene activity during development. *Science*, **187**, 226–32.

Jeppesen, P. & Turner, B.M. (1993) The inactive X-chromosome in female mammals is distinguished by a lack of histone H4 acetylation, a cytogenetic marker for gene expression. *Cell*, **74**, 281–9.

Johnston, M. (1987) A model fungal gene regulatory system: the GAL genes of *Saccharomyces cerevisiae*. *Microbiol Rev*, **51**, 458–76.

Krumlauf, R. (1993) *Hox* genes and pattern formation in

the brachial region of the vertebrate head. *Trends Genet*, **9**, 106–12.

Lawrence, P.A. (1992) *The Making of a Fly*. Blackwell Scientific Publications, Oxford.

Le Mouellic, H., Lallemand, Y. & Brulet, P. (1992) Homeosis in the mouse induced by a null mutation in the *Hox-3.1* gene. *Cell*, **69**, 251–64.

Lyon, M. (1993) Epigenetic inheritance in mammals. *Trends Genet*, **9**, 123–7.

McGinnis, W. & Krumlauf, R. (1992) Homeobox genes and axial patterning. *Cell*, **68**, 283–302.

Messmer, S., Franke, A. & Paro, R. (1992) Analysis of the functional role of the Polycomb chromodomain in *Drosophila melanogaster*. *Genes Dev*, **6**, 1241–54.

Morata, G. & Lawrence, P.A. (1977) Homeotic genes, compartments and cell determination in *Drosophila*. *Nature*, **265**, 211–16.

Pankratz, M.J. & Jäckle, H. (1990) Making stripes in the *Drosophila* embryo. *Trends Genet*, **6**, 287–92.

Paro, R. (1990) Imprinting a determined state into the chromatin of *Drosophila*. *Trends Genet*, **6**, 110–14.

Reuter, G. & Spierer, P. (1992) Position effect variegation and chromatin proteins. *BioEssays*, **14**, 605–12.

Riggs, A.D. & Pfeifer, G.D. (1992) X-chromosome inactivation and cell memory. *Trends Genet*, **8**, 169–74.

Rothe, M., Pehl, M., Taubert, H. & Jäckle, H. (1992) Loss of gene function through rapid mitotic cycles in the *Drosophila* embryo. *Nature*, **359**, 156–9.

Stanojevic, D., Small, D. & Levine, M. (1991) Regulation of a segmentation stripe by overlapping gradients of transcriptional activators and repressors in the *Drosophila* embryo. *Science*, **254**, 1385–7.

St. Johnston, D. & Nusslein-Volhard, C. (1992) The origin of pattern and polarity in the *Drosophila* embryo (Review). *Cell*, **68**, 201–19.

Surani, M.A. (1991) Genomic imprinting: developmental significance and molecular mechanism. *Curr Opin Genet Dev*, **1**, 241–6.

Turner, B.M., Birley, A.J. & Lavender, J. (1992) Histone isoforms acetylated at specific lysine residues define individual chromosomes and chromatin domains in *Drosophila* polytene nuclei. *Cell*, **69**, 375–84.

Wu, L. & Russell, P. (1993) Nim1 kinase promotes mitosis by inactivating Wee1 tyrosine kinase. *Nature*, **363**, 738–41.

INDEX